Sulphur Dioxide and Nitrogen Oxides in Industrial Waste Gases: Emission, Legislation and Abatement

Edited by

Daniel van Velzen

*Commission of the European Communities,
Joint Research Centre,
Environment Institute,
Ispra, Italy*

SPRINGER-SCIENCE+BUSINESS MEDIA, B.V.

Based on the lectures given during the Eurocourse on
Sulphur Dioxide and Nitrogen Oxides in Industrial Waste Gases:
Emission, Legislation and Abatement
held at the Joint Research Centre Ispra, Italy, September 3–7, 1990

ISBN 978-94-010-5608-3   ISBN 978-94-011-3624-2 (eBook)
DOI 10.1007/978-94-011-3624-2

Publication arrangements by
Commission of the European Communities
Directorate-General Telecommunications, Information Industries and Innovation,
Scientific and Technical Communication Unit, Luxembourg

EUR 13599
© 1991 Springer Science+Business Media Dordrecht
Originally published by Kluwer Academic Publishers in 1991
LEGAL NOTICE
Neither the Commission of the European Communities nor any person acting on behalf of the
Commission is responsible for the use which might be made of the following information.

Printed on acid-free paper

All Rights Reserved
No part of the material protected by this copyright notice may be reproduced or
utilized in any form or by any means, electronic or mechanical,
including photocopying, recording or by any information storage and
retrieval system, without written permission from the copyright owner.

EURO
COURSES

A series devoted to the publication of courses and educational seminars organized by the Joint Research Centre Ispra, as part of its education and training program.
Published for the Commission of the European Communities, Directorate-General Telecommunications, Information Industries and Innovation, Scientific and Technical Communications Service.

The EUROCOURSES consist of the following subseries:

- Advanced Scientific Techniques
- Chemical and Environmental Science
- Energy Systems and Technology
- Environmental Impact Assessment
- Health Physics and Radiation Protection
- Computer and Information Science
- Mechanical and Materials Science
- Nuclear Science and Technology
- Reliability and Risk Analysis
- Remote Sensing
- Technological Innovation

CHEMICAL AND ENVIRONMENTAL SCIENCE

Volume 3

The publisher will accept continuation orders for this series which may be cancelled at any time and which provide for automatic billing and shipping of each title in the series upon publication. Please write for details.

Sulphur Dioxide and Nitrogen Oxides
in Industrial Waste Gases:
Emission, Legislation and Abatement

CONTENTS

	List of Contributors	vii
1.	PREFACE AND COURSE INTRODUCTION *D. van Velzen*	1
2.	EMISSION SOURCES AND QUANTITIES *B. Heinen*	9
3.	ATMOSPHERIC CHEMISTRY OF SULPHUR DIOXIDE AND NITROGEN OXIDES *G. Restelli and H. Stangl*	23
4.	DISPERSION AND TRANSPORT OF ATMOSPHERIC POLLUTANTS *S. Cieslik*	47
5.	CONTROL OF AIR POLLUTION - THE EUROPEAN COMMUNITY APPROACH *H.P. Stief-Tauch*	73
6.	DESULPHURIZATION OF FLUE GASES ON THE BASIS OF LIME OR LIMESTONE SCRUBBING *G. Mittelbach*	93
7.	THE WELLMAN LORD PROCESS *U. Neumann*	111
8.	ISPRA MARK 13A DESULPHURIZATION PROCESS *H. Langenkamp*	139
9.	BF / UHDE / MITSUI-ACTIVE COKE PROCESS FOR SIMULTANEOUS SO_2- AND NO_X-REMOVAL *E. Richter*	157
10.	WALTHER PROCESS *W. Schulte*	167
11.	EBDS-PROCESS *H.-R. Paur*	183
12.	PRIMARY MEASURES FOR NO_X REDUCTION *H.G. Bos*	205
13.	HIGH AND LOW DUST SCR PROCESSES *E. Weber and D. Schmidt*	223
14.	COSTS OF DESULPHURIZATION AND DENOXING *D. van Velzen*	235
15.	SITUATION IN THE UNITED STATES AND JAPAN *G. Caprioglio*	263

LIST OF CONTRIBUTORS

H. G. BOS,
Stork Boilers,
P.O. Box 20,
NL-7550 CB Hengelo, THE NETHERLANDS

G. CAPRIOGLIO
Ferlini / General Atomics Development Corporation,
P.O. Box 85608,
San Diego, California, UNITED STATES OF AMERICA

S. CIESLIK
Commission of the European Communities,
Joint Research Centre (JRC), Environment Institute,
I-21020 Ispra, ITALY

B. HEINEN
Energie Consulting Heidelberg,
Im Breitspiel 7,
D-6900 Heidelberg, Federal Republic of GERMANY

H. LANGENKAMP
Commission of the European Communities,
Joint Research Centre (JRC), Environment Institute,
I-21020 Ispra, ITALY

G. MITTELBACH
Deutsche Babcock Anlagen AG,
Postfach 4 + 6,
D-4150 Krefeld, Federal Republic of GERMANY

U. NEUMANN
Davy McKee AG,
Borsigallee 1 - 7,
D-6000 Frankfurt / Main 60, Federal Republic of GERMANY

H.-R. PAUR,
Kernforschungszentrum Karlsruhe GmbH,
Laboratorium für Aerosolphysik und Filtertechnik I,
Postfach 3640,
D-7500 Karlsruhe 1, Federal Republic of GERMANY

G. RESTELLI
Commission of the European Communities,
Joint Research Centre (JRC), Environment Institute - Air Chemistry,
I-21020 Ispra, ITALY

E. RICHTER,
DMT Gesellschaft für Forschung und Prüfung mbH,
Institut für Wärme und Stromerzeugung,
Franz Fischer-Weg 61,
D-4300 Essen 13, Federal Republic of GERMANY

D. SCHMIDT,
Institut für Umweltsverfahrenstechnik, Universität Essen,
Leimkugelstrasse 10,
D-4300 Essen 13, Federal Republic of GERMANY

W. SCHULTE,
Krupp Koppers GmbH,
POB 10 22 51,
D-4300 Essen 1, Federal Republic of GERMANY

H. STANGL
Commission of the European Communities,
Joint Research Centre (JRC), Environment Institute - Air Chemistry,
I-21020 Ispra, ITALY

H.P. STIEF-TAUCH
Commission of the European Communities,
Directorate General XI - B,
Rue de la Loi 200,
B-1049 Brussels, BELGIUM

D. VAN VELZEN
Commission of the European Communities,
Joint Research Centre (JRC), Environment Institute,
I-21020 Ispra, ITALY

E. WEBER
Institut für Umweltsverfahrenstechnik, Universität Essen,
Leimkugelstrasse 10,
D-4300 Essen 13, Federal Republic of GERMANY

PREFACE AND COURSE INTRODUCTION

D. VAN VELZEN
Commission of the European Communities,
Joint Research Centre Ispra,
Environment Institute,
I-21020 Ispra (Varese) ITALY

1. Introduction

Worldwide, there is an ever increasing interest and concern about the destructive effects of air pollution on man's ecosystem. The growing awareness of these effects has revealed the need to take adequate measures to minimize the emission of air polluting products. The two most important contaminants, occurring in the largest concentrations and quantities, are sulphur dioxide and nitrogen oxides. Both pollutants are formed mainly during the combustion of fossil fuels, particularly by power stations and traffic. The effects of air pollution caused by these two contaminants have already been studied for several decades and measures to protect the environment against their adverse effects are now operative in many countries.

The present volume contains the proceedings of a Eurocourse held in Ispra in September 1990. The course was meant to give an overview of present knowledge concerning the emission sources and quantities, to cover features of present legislation and to give a survey of the most important modern abatement techniques for SO_2 and NO_x. It was mainly addressed to higher and medium management in the power, chemical and similar industries, particularly from those countries where the fight against air pollution is still in its infancy.
   Obviously, it was not possible to cover completely the whole range of subjects during the limited duration of a Eurocourse. For every facet of the problem, choices had to be made. A short justification of these choices is given in the following paragraphs.

2. Emission and transport

This section consists of a set of three papers. The first one *(Heinen)* gives a survey of the sources and quantities of sulphur dioxide and nitrogen oxides emissions. The chemical processes which take place in the atmosphere and which are responsable for the conversion of the gaseous products into the acidic precipitates, causing the destructive effects known as "acid rain", are discussed in the second paper *(Restelli and Stangl)*.

The third paper *(Cieslik)* presents the state of the art in atmospheric modelling used for the prediction of the airborne transport of the contaminants. These three review papers give together a very satisfactory survey of the present state of knowledge in the field of the emission and transport of air pollutants.

3. Legislation

Limitations of SO_2 and NO_x emissions are imposed by national and international regulations. Currently there are approximately 20 countries where national emission standards are operative. Recently important developments in control regulations took place, both nationally and internationally.
One of the most important international agreements is the United Nations Economic Commission for Europe (UNECE) Convention on Long-range Transboundary Air Pollution, signed in 1979 in Geneva by 33 countries. Later, in 1985, this was followed by the adoption of a protocol to the convention which called for a 30% reduction in total SO_2 emission by 1993, compared with 1980 levels. This protocol was signed by 21 countries, which have become known as the "30% club".

The EC Directive on the limitation of emissions of pollutants from large combustion plants was agreed in 1988 and applies to the twelve member states of the European Community. The target for existing plants is a 23% reduction in total EC emissions by 1993 and a 57% reduction by 2003. The baseline figure is, here also, the emission rate for 1980.
The reduction targets differ from country to country. Less developed countries like Greece, Portugal and Ireland are even allowed to increase their emissions, whereas the most stringent targets (70%) are set for Belgium, Germany, France and the Netherlands.
The background and details of the approach of the European Community in matters of legislation to prevent air pollution are discussed extensively in the fourth paper of the course *(Stief-Tauch)*

4. Sulphur dioxide abatement

The regulations for sulphur dioxide emissions have become considerably more stringent with time, especially during the 1980s. This has led to the application of various measures to reduce the emission of SO_2. There are a number of options to realize these reductions:
  i) -   decreasing the sulphur content of the fuel or switching to other fuels;
  ii) - the use of processes to decrease emissions during combustion;
  iii) - the application of flue gas desulphurisation (FGD) processes.

There are several alternatives in each category, often resulting in different levels of emission of SO_2. Particularly in the field of FGD processes, the alternatives are manifold. It is in this field that rigorous choices had to be made concerning the contents of this course.

4.1 ACTIONS CONCERNING THE FUEL

It is very common practice in countries where regulations for SO_2 emission are applied for the first time to <u>switch fuels</u>. For coal fired power stations, the change from high sulphur to low sulphur coal is an obvious one, and also in many other situations switching from high sulphur fuel oil to natural gas fired furnaces is applied. Such solutions usually require a limited investment compared to the installation of bulky FGD equipment and the additional necessary particulate removal systems. However, availability and price of the fuels set a limit to the applicability of this alternative for the reduction of SO_2-emissions.

The second alternative in this category is the removal of sulphur from the fuel before combustion. The _desulphurization of coal_ has been studied extensively, but, until now, no practical industrial process has been developed. The main reason is that by washing and/or froth flotation of the coal a certain degree of sulphur removal can be obtained, but that removal of the organically bound sulphur and pyrites is as yet unpracticable. Studies into the microbiological desulphurization of coal are at present under way, but are still at the laboratory stage and the projected process costs seem to be high.

It follows that switching fuels is a rather obvious alternative, whereas the desulphurization of fuels is as yet not practicable. For these reasons, no paper dealing with the desulphurization of coal or actions concerning the fuel was included in the course.

4.2 ACTIONS CONCERNING COMBUSTION

Combustion technologies exist where the control of the emission is an integral part of the process design. Examples include the _fluidised bed combustion_ of coal and the application of sorbent injection. In both cases the intrinsically formed sulphur dioxide reacts with and is bound to an active solid material, usually calcium or magnesium oxide.

Research on fluidised bed combustion (FBC) for energy production was started in the early 1960s. In this process, the fuel is burned in a furnace containing a bed of finely-divided solid particles of high melting point. The combustion air passes upwards through the bed and causes fluidization of the particles. The total mass behaves like a boiling fluid. Typically, FBC operates at a temperature of 850°C, which is several hundred degrees lower than conventional combustion. With this lower combustion temperature, the formation of nitrogen oxides is largely avoided. The addition of a calcium or magnesium mineral like dolomite to the solids can to a certain extent prevent the emission of sulphur dioxide. The formed SO_2 is bound to the solids in the form of sulphate or sulphite and leaves the process together with the spent solids, coal ash and unconverted sorbent. Most of the early FBC units experienced major mechanical and material problems, including erosion of the heat exchange tubing and plugging of coal feed systems. Nowadays most of these problems have been solved and coal fired FBC boiler systems are fairly widespread in Western Europe and North America, with over 200 industrial size units in operation in utilities and process industries with a capacity of up to approximately 100 MWe.

The disadvantages of the method are its high cost, the fact that it is only applicable to new boilers of relatively small size and the high consumption of sorbent material needed to achieve at a satisfactory degree of desulphurization. In our opinion, the contribution of FBC to the solution of the problem of SO_2 emission is small, so that no detailed specific paper was included in the Eurocourse.

Sorbent injection into the boiler, flue gas duct or a combination of both has also been under study for a long time. A dry calcium or sodium based solid sorbent is used for this purpose. The process produces a dry sodium or calcium sulphite/sulphate waste mixed with the fly ash. Generally, countries starting with flue gas desulphurisation activities consider this approach interesting. It involves low capital costs and is easy to install. However, the degree of desulphurization obtainable is low, typically of the order of 50%. In general it can be said that sorbent injection is unable to achieve exit SO_2 concentrations in compliance with modern regulations (200 - 400 mg/m^3).
For these reasons this subject was not treated separately in the Eurocourse.

4.3 FLUE GAS DESULPHURIZATION (FGD)

The most widely used way of reducing the emission of sulphur dioxide is flue gas desulphurization (FGD). The technique consists basically of contacting the flue gas with a liquid or slurry containing a reactant for SO_2 in specially designed reactors. The vast majority of the FGD systems presently in use involve *wet scrubbing*. The process is based on a gas/liquid reaction where the liquid sorbent is injected into the flue gases, usually in an open tower. Here also, the application of calcium based reactants, lime or limestone, is wide-spread. Normally, an oxidation step is incorporated to ensure the production of calcium sulphate (gypsum). At present, processes producing gypsum of a saleable quality are highly favoured.

There are a large number of variations based on *lime or limestone scrubbing*. The differences between the various processes involve the use of additional chemicals, like formic acid, adipic acid etc. or other ways of maintaining a constant pH. There are so many process variations on the market that not all can be mentioned separately. For a complete list and description of available processes, the reader interested is referred to specialized literature, like the FGD handbook prepared by IEA [1]. A representative example of a lime/limestone scrubbing process is described in detail in the course *(Mittelbach)*.

Lime/limestone scrubbing can also be carried out in so-called *spray dry scrubbers*. They have been developed as an alternative, especially for smaller installations and for lower SO_2-concentrations. Here, the sorbent slurry is sprayed into the flue gas and subsequently dried using the sensible heat of the flue gas. The product is generally a free-flowing fine solid sulphite/sulphate mixture, inevitably containing a proportion of unreacted sorbent.

The large number of lime/limestone scrubbing plants in operation or to be installed in the near future will produce a very large amount of gypsum. This might create big problems in marketing and/or in waste disposal, which is an important reason for the development of alternative processes, the so-called *regenerable processes*. These are processes where the sorbent for SO_2 can be recovered after a regeneration step. In many cases concentrated sulphur dioxide is the end product, which can subsequently be converted into sulphuric acid or pure sulphur. There are also processes where sulphuric acid or ammonium sulphate is produced directly.

The most widely used regenerable FGD system is the *Wellman-Lord process*, which uses a sodium sulphite solution as the sorbent, making use of the sulphite/bisulphite equilibrium for the regeneration. Operational experience with this process is given in the course *(Neumann)*. Another important example of a regenerable process is the *Walther process*, based on the neutralization of SO_2 with ammonia *(Schulte)*. As an example of a promising new regenerable process under development, the course also contains a paper on the *Ispra Mark 13A process*. This process is based on the reaction of SO_2 with bromine and the electrolytic regeneration of the sorbent *(Langenkamp)*.

The choice of the three regenerable processes was mainly based on to the differences in their operating principles. There are many other regenerable processes in various states of development. A rather complete survey of these processes is included in Ref. [1].

4.4 IMPORTANCE OF FGD

The first FGD installations were installed in the early 1970s in Japan and the USA, followed by a major expansion in Europe during the late 1980s, particularly in Germany. By the end of 1990 about 540 boilers were equipped with FGD plants, equivalent to a total installed capacity of approximately 147 GWe.

The IEA Coal Research keeps a data-base of all known installations [2]. A survey of the amount of installed FGD capacity in different countries, divided into various process categories is given in Table 1. The figures illustrate very well the importance of FGD as a means of reducing SO_2 emissions compared with other methods such as FBC etc.

It is believed that another big international effort must be undertaken in the near future. It is predicted that by the end of the century, the total global installed capacity of FGD will probably be increased to 260 - 280 GWe [3].

5. NOx abatement

In the field of the reduction of nitrogen oxides emissions, the number of available technologies is less extensive than for the removal of sulphur dioxide. In practice, the best and most economic approach to the problem for stationary sources, is the application of combustion modification, the so-called *primary measures*. A careful application of these measures can already reduce the NO_x emissions by 60% at minimal cost *(Bos)*.

The other alternative is the application of the process of *selective catalytic reduction (SCR)*, where NO_x is catalytically reduced to N_2 and H_2O by reacting with NH_3. The reaction takes place typically at a temperature of 300 - 400°C *(Weber and Schmidt)*.
In practice, there are no other alternatives of any industrial importance, other than the processes mentioned in the following section.

6. Combined SO_2/NO_x removal

There are a number of processes where the removal of sulphur dioxide is combined with the denoxing of the flue gas. The best known process in this category is the *activated carbon process* where SO_2 is absorbed on activated carbon, followed by the injection of ammonia before the flue gas passes into a second stage of the reactor. In the second stage at low temperature NO_x is then reduced by reaction with NH_3 to N_2 and H_2O *(Richter)*.

Countries	Sorbent injection	Spray dry scrubbers	Lime-stone gypsum	Other wet processes	Regenerable processes	Not known	Total
Austria	315	755	690	100			1 860
Canada	300						300
China				700			700
Denmark		700	500			370	1 570
Germany	760	2 920	35 150	800	1 770	470	41 870
Finland	280	260					540
France	600						600
Italy					30		30
Japan			13 680	200			13 880
Netherlands			2 730				2 730
Sweden	530	550		10			1 090
Turkey			340				340
USA	550	7 600	6 510	57 970	3 010	6 090	81 730
USSR	45						45
TOTAL	3 380	12 785	59 600	59 780	4 810	6 930	147 285

TABLE 1
Installed FGD capacity (MWe) by the end of 1990 [3]

There are other combined SO_2/NO_x removal processes, for instance the catalytic gas phase oxidation of SO_2 to H_2SO_4, followed by selective catalytic reduction of the nitrogen oxides (the German DESONO$_X$ and the Danish SNOX process). The NOx removal step in these processes consists of a "classical" SCR process.

Another elegant process proposal involves the use of radical formation by the acceleration of electrons, the EBDS process *(Paur)*. A paper on cost estimation *(van Velzen)* and a review of the situation in Japan and the U.S.A. *(Caprioglio)* complete the contents of the course.

7. References

[1] Klingspor J. and Cope D., 1987, "FGD handbook, flue gas desulphurisation systems", ICEAS/BS, IEA Coal Research London UK, 271 pp

[2] IEA Coal Research, 1991, IEA Coal Research FGD and NOx control installations database, London UK, IEA Coal Research

[3] A. Hjalmarsson, "FGD installations on coal-fired plants", Symposium Desulphurisation 2 - Technologies and Strategies for Reducing Sulphur Emissions, Sheffield, March 1991

EMISSION SOURCES AND QUANTITIES

BERND HEINEN
Energie Consulting Heidelberg
Im Breitspiel 7
D-6900 HEIDELBERG
Federal Republic of GERMANY

1. Introduction

The growing industries of the last two centuries have effected an increase of pollutant emissions like SO_2 and NO_x. Today, these manmade emissions make worldwide up for about 50% of the total of natural and anthropogenic sulphur and nitrogen oxides released into the atmosphere [1].

Once dispersed, SO_2 and NO_x can directly affect nature and man's health. Furthermore, chemical reactions (Fig. 1) will produce e.g. ozone which is a precursor of smog, or sulphuric and nitric acid which are responsible for acid rain and its negative effects on materials, soil, water and other ecosystems.

As an example it is noted that, together with other pollutants, SO_2 and NO_x are to to blame for forest damage and the dying of fish in the acidified lakes of Scandinavia.

The acidified lakes in Scandinavia indicate another problem: high winds transport the pollutants away from the emission source and cause them to settle over a wide and sometimes remote area.
In this way, Norway gets for example about 92% of the deposited sulphur from foreign countries, Sweden about 82% [2]. In both countries, the acidified lakes form an immense problem.

2. Emission Sources

As already mentioned, there are two principal sources for pollutants such as SO_2 and NO_x, i.e. natural and manmade sources.

Natural SO_2 is, as a general rule, released from volcanic sources and to a much lower extent from marsh gases. In nature NO_x is mainly produced in the course of the chemical and bacterial denitrification processes going on in the soil.

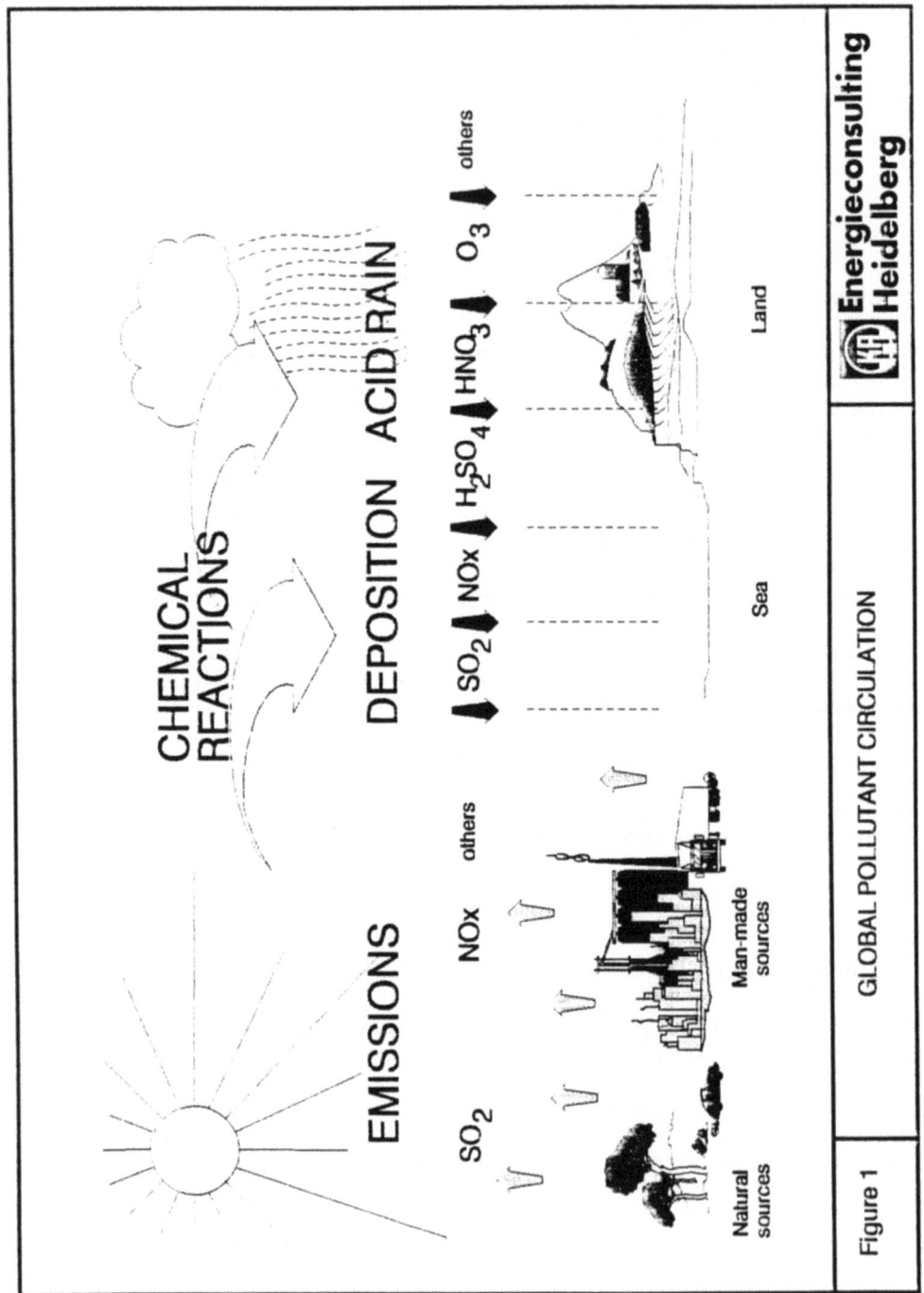

Figure 1 — GLOBAL POLLUTANT CIRCULATION

On the other side, manmade pollutants are, above all, produced in combustion processes, where coal, oil, gas and biomass are used as fuel for heat and power generation, or where gasoline is used in internal combustion engines.

2.1. Mechanism of Pollutant Development

Depending on the nature of the fuel, there are different steps in the combustion process. These are for example for solid fuels:

- heating
- drying
- gasification
- combustion

The time required for the entire combustion process is in general largely determined by the water content of the fuel. The various above mentioned process steps are all proceeding simultaneously.
The sulphur and nitrogen contained in the fuel and released during the gasification process, are oxidized during the combustion process and converted into SO_2 and NO_x.

A small part of the sulphur (up to 5%) is converted into SO_3 which is the precursor of sulphuric acid [3]. Most of the sulphur is released into the atmosphere in the form of gaseous sulphur oxides. In certain coal and oil combustion processes, it is possible to retain up to 40% of the sulphur in the fly ash or dust [3].

Most of the nitrogen oxides produced during combustion processes consist of nitrogen oxide (NO) and to a smaller extent (up to 5%) of nitrogen dioxide (NO_2) [4]. Once released into the atmosphere, NO is being oxidized to NO_2.

Besides the NO_x produced as a result of the nitrogen contained in the fuel ('fuel NO_x') there are two other processes of NO_x formation from the air used in the combustion. They are:

- the 'thermal NO':
  Nitrogen contained in the combustion air is converted into NO_x at high temperatures (above 1300 °C), following the 'Zeldovic'- mechanism [4];

- the 'prompt NO'
  is generated by the reaction of fuel radicals with nitrogen of the combustion air. The portion of the prompt NO_x in the final NO_x concentration is of minor significance.

Thus, contrary to SO_2 development, the formation of NO_x is not merely attributable to the nitrogen content in the fuel, but depends highly on furnace conditions:

High temperatures and a high oxygen content, especially in or near the flame, favour NO_x development. That is why it is possible to reduce NO_x emissions significantly by modifying the combustion conditions in the furnace ('primary measures').

2.2. Manmade Emission Sources

There are four major categories which are to blame for SO_2 and NO_x emissions:

- industry
- traffic
- power plants
- domestic and other small sources

In European countries the NO_x-emissions from traffic ('the mobile sources') typically account for more than 50% of the total NO_x emissions [5].
However, SO_2 emissions from mobile sources account for less than 10% in most countries, depending on the fuels used in the combustion processes for domestic heating and for electric power production.

Fig. 2 gives as an example the distribution of the SO_2 and NO_x emission sources in the Federal Republic of Germany in 1985. Power plants make up for 63.5% of total SO_2 emissions, the industry for 23.2%, domestic and other small sources for 9.5% and traffic for 3.8%.
The NO_x emission sources are distributed over traffic (58.8%), power plants (26%), industry (10.5%), domestic and other small sources (4.7%).

Each one of the four principal sources can be split up into some subsections, detailed information being given in the following.

2.2.1. Industry
Emissions from industry vary extremely in composition and their volume, depending on the type of the respective industrial process. Major emission sources are:

- Oil industry
The oil industry processes crude oil to gasoline, bitumen, lube-oil, and other raw materials for the petrochemical industry.
SO_2 and NO_x are produced in various thermal and combustion processes.

A refinery typically consumes about 5 to 8% of its oil input for production, heating etc. [6]. The emissions depend on the size of the plant, the sulphur content of the oil and the fuel used in the processes.

- Chemical industry
Main sources for SO_2 and NO_x emissions are sulphuric and nitric acid production plants. Sulphuric acid is produced by absorption of SO_3 in water. Sulphur trioxide itself is produced by catalytic oxidation of SO_2, which on its turn is produced by the combustion of sulphur or by roasting of sulphur containing ores (pyrites).

of operation, and fuel used. When diesel fuel is burnt, this contributes to SO_2 emissions. Other pollutants fuel from traffic are carbon monoxide and hydrocarbons.

2.2.3. Power plants
Electrical power generation is responsable for the production of immensely largest flue gas volumes. Quantitatively, this is the largest source of air pollution of the four mentioned items.

Fossil fuels are burnt in the combustion processes and produce NO_x and, depending on the type of fuel, also SO_2, CO, dust, chlorine and fluor.

Furnaces firing gas and oil in general convert all the sulphur contained in the fuel into SO_2. However, when the mineral coal or lignite are fired, it is possible to retain 5% and even up to 70% of the sulphur in the ash [8], respectively.

Crude oil can contain up to 5% sulphur, whereas the typical sulphur content of fuel oil is in Europe about 0.3%.
A typical German mineral coal contains about 1% of sulphur, and the sulphur content of lignite is about 2%.

The NO_x-emissions largely depend on the furnace design. Full load operation and high temperatures like in wet-bottom boilers favor NO_x emissions.

2.2.4. Domestic and Others
The principal energy demand in the domestic sector and others e.g. small trade and business as well as military facilities, is for room heating. The most frequently used fuels are fuel oil and natural gas. Both produce NO_x, whereas SO_2 is primarily produced during the combustion of domestic fuel oil.

3. Emission Quantities

3.1. Definition
It is possible to come to a quantitative definition of the emissions, by monitoring the respective emission sources. Generally, only some of the sources, primarily larger power plants, are being monitored i.e. their volume flow and pollutants like SO_2 and NO_x. Thus, all other sources must therefore be defined theoretically:

By using emission factors that are related to specific production data, i.e. the fuel, energy and raw material consumption, it is possible to calculate the emissions. There is no doubt that these calculations are subject to certain uncertainties. Nevertheless, today they are, indeed, the most accurate basis for emission inventories. Table I shows as example the emission factors for some combustion facilities.

	hard coal	lignite	pressing lignite	fuel oil	natural gas
SO_2	19 x S	10 x S	10 x S	20 x S	20 x S
NO_x (as NO_2)	1.5 - 3.0	0.4 - 0.8	0.96	5.3	3.0 - 5.0

TABLE I
Emission factors (g/kg fuel) for the combustion of different fuels [8] (S in %).

The application of the emission factors is elucidated by an example:
From Table 1 it follows that for natural gas the emission factor is 20 x S, where S is correlated to the sulphur content of the fuel (in wt%). Assuming that the S-content of natural gas is approximately 0.03 wt%, it follows that the combustion of 1 kg of natural gas yields 0.60 g of SO_2.

3.2. Emission quantities in Europe

Table 2 and 3 show the SO_2 and NO_x emission quantities in Europe for 1985 [9]. The total SO_2-emissions amount to 45.6 million per year. The total NO_x-emissions are 18.7 million tons per year. In comparison, the worldwide natural and manmade emissions are estimated to be about 400 million tons of SO_2 and about 175 million tons of NO_x per year [10].

Major SO_2 emitters in Europe are the URSS with 11 million tons; equivalent to 24%, the former German Democratic Republic (GDR) with 5 million tons (11%), and Poland with 4.3 million tons (9%).
Referring to the NO_x-emissions (Table 3), the major emitters are the URSS and the Federal Republic of Germany (FRG) with 2.9 million tons or 16% each, France and Great Britain follow with 1.69 million tons (9%) each.

Additionally, the tables show in the last three columns specifically the emission quantities, related to the energy consumption, surface and also per inhabitant. Generally, countries using high-sulphur fuel and not providing extensive desulphuration measures have the highest specific SO_2 emissions. As to the emissions related to the energy consumption, this concerns above all countries like Spain and most of the countries in Eastern Europe, such as the GDR, Czechoslovakia and Hungary. On a per inhabitant basis, the SO_2 emissions in East Europe countries are highest. It can be assumed that compared to countries in Western Europe, in the first place the lower power plant efficiency in Eastern countries is to be blamed.

3.3. Emission forecasts

SO_2 and NO_x emission inventories for the future have to consider a lot of factors like energy supply, fuel used, sulphur content of the fuel, technical

Country	Inhabitant	fossil fuel c.	SO$_2$-Emissions 1985			
	million	$\frac{GJ}{inhab.}$	$\frac{1000 \text{ tons}}{a}$	$\frac{kg}{a \times TJ}$	$\frac{t}{a \times km^2}$	$\frac{kg}{a \times inhabit.}$
Albany	3,0	36	50	463	1,7	17
Austria	7,6	101	170	222	2,0	22
Belgium	9,9	134	468	353	15,3	47
Bulgaria	9,0	162	1140	786	10,3	127
Czechoslovakia	15,5	179	3150	1136	24,6	203
Danmark	5,1	156	326	408	7,6	64
Finland	4,9	127	370	592	1,1	75
France	55,3	100	1846	334	3,4	33
FRG	61,0	160	2440	251	9,8	40
GDR	16,6	227	5000	1325	46,2	301
Greatbritain	56,1	140	3540	449	14,5	63
Greece	9,9	67	360	544	2,7	36
Hungary	10,6	108	1420	1227	15,3	133
Ireland	3,5	91	138	427	2,0	39
Iceland	0,24	87	6	286	0,1	25
Italy	57,1	92	3150	549	10,5	55
Jugoslawia	23,1	69	1800	1125	7,0	78
Luxemburg	0,37	301	14	127	5,4	38
Netherlands	14,5	167	316	131	7,7	22
Norway	4,2	102	100	237	0,5	24
Poland	37,2	136	4300	852	13,8	116
Portugal	10,2	34	306	892	3,4	30
Rumania	23,0	138	200	63	0,8	9
Spain	39	58	3250	1441	6,4	83
Sweden	8,4	90	272	362	0,6	33
Switzerland	6,4	87	96	173	2,3	15
Turkey	49,3	26	322	251	0,4	7
USSR (europ)	195,3		11100		3,3	57
Europe	736,3		45650			

TABLE 2: SO$_2$ Emissions in Europe [9]

Country	Inhabitant	fossil fuel c.	NOx-Emissions (as NO2)			
	million	GJ/inhabit.	year	1000 tons/a	kg/(a×TJ)	t/(a×km²)	kg/(a×inhab.)
Albany	3,0	36					
Austria	7,6	101	1985	216	282	2,6	29
Belgium	9,9	134	1984	385	290	12,6	39
Bulgaria	9,0	162		200	138	1,8	22
Czechosl.	15,5	179	1985	1120	404	8,8	72
Danmark	5,1	156	1985	290	363	6,7	57
Finland	4,9	127	1983	250	463	0,7	51
France	55,3	100	1985	1693	306	3,1	31
FRG	61,0	160	1985	2930	301	11,8	48
GDR	16,6	227	1986	955	253	8,8	58
Greatbr.	56,1	140	1984	1690	224	6,9	30
Greece	9,9	67		150	227	1,1	15
Hungary	10,6	108	1985	300	259	3,2	28
Ireland	3,5	91	1984	75	223	1,1	21
Iceland	0,24	87		10	476	0,1	41
Italy	57,1	92	1983	1462	302	4,9	26
Jugosl.	23,1	69		190	119	0,7	8
Luxemb.	0,37	301	1985	22	200	8,5	60
Netherl.	14,5	167	1983	480	175	11,8	33
Norway	4,2	102	1984	138	309	0,7	33
Poland	37,2	136		840	166	2,7	23
Portugal	10,2	34	1983	330	932	3,7	32
Rumania	23,0	138		390	123	1,6	17
Spain	39	58	1983	950	408	1,9	24
Sweden	8,4	90	1984	289	406	0,6	35
Switzerl.	6,4	87	1984	214	396	5,2	34
Turkey	49,3	26		175	137	0,2	4
USSR (eu)	195,3		1985	2930		0,9	15
Europe	736,3			18674			

TABLE 3: NOx Emissions in Europe [9]

	Year	Amount	1980	Year	Amount	Reduction Rate from	Reduction Rate to	%	Reduction Rate from 1980 to	%
BELGIUM	1970	1 033	799	1984	467	1970	1984	55	1984	42
DENMARK	1970	523	438	1985	326	1970	1985	38	1985	26
FRANCE[1]	1971	2 966	3 558	1985	1 845	1971	1985	38	1985	48
GERMANY, F.R.[2]	1970	3 600	3 200	1985	2 400	1970	1985	33	1985	25
GREECE			800*	1983	720*				1983	10
IRELAND	1972	186	219	1985	138	1972	1985	26	1985	37
ITALY	1972	3 200	3 800	1985	2 503	1972	1985	22	1985	34
LUXEMBOURG	1970	35	23	1985	13	1970	1985	63	1985	44
NETHERLANDS[3]	1970	685	487	1985	275	1970	1985	60	1985	44
PORTUGAL	1970	116	266	1983	305	1970	1983	+ 163	1983	+ 15
SPAIN	1975	3 004	3 250	1983	3 250	1975	1983	+ 8	1983	0
UNITED KINGDOM[4]	1970	6 120	4 670	1985	3 580	1970	1985	42	1985	23

1) Total Emission 1987 (preliminary figures): 1,520 000 t/a
2) "   "   1987   "   "   : 2,000 000 t/a
3) "   "   1987   "   "   :   282 000 t/a
4) "   "   1987   "   "   : 3,680 000 t/a

Table 4: Development of the SO_2-emissions in Europe
Source: (11)

* According to ECE from 1988, the total SO_2-emissions amounts to 400 kt in 1980 and to 360 kt in 1983. If these figures are correct Greece would have a better position in the ranking list.

developments and regulation for emission control. Probably, single cou
will come to a significant change in their emissions. E.g. the Federal Re
of Germany is expected to experience a SO_2/NOx emission reduction of
50% and 35% respectively between 1986 and 1988 [9].

However, investigations carried out for the European Economic Com
[1] report a change in SO_2 emissions of -15 up to + 2% when compar
year 2000 and 1980. As to the NO_x, an increase of emissions is to be e
as a result of more traffic in the Community.

In this context it is interesting to compare the development of the
NO_x emissions in the countries of the European Community over t
1970 to 1985 with an intermediate balance taken at 1980. These
given in Tables 4 and 5.

It follows that, as a general rule, the NO_x emissions strongly increas
the years 1970-1980, and in the period from 1980 to 1985 tended
out. An explanation of this phenomenen is not given by the auth
reference.
In the case of the emission of SO_2, the picture is different. Here,
decrease can be observed. The average reduction rates over the p
1980 to 1985 is about 40%.

It is not at all clear how the predictions of the pollutant emissions
to 2000 will prove to be realistic. The present situation and future
are largely uncertain.

developments and regulation for emission control. Probably, single countries will come to a significant change in their emissions. E.g. the Federal Republic of Germany is expected to experience a SO_2/NOx emission reduction of about 50% and 35% respectively between 1986 and 1988 [9].

However, investigations carried out for the European Economic Community [1] report a change in SO_2 emissions of -15 up to + 2% when comparing the year 2000 and 1980. As to the NO_x, an increase of emissions is to be expected as a result of more traffic in the Community.

In this context it is interesting to compare the development of the SO_2 and NO_x emissions in the countries of the European Community over the years 1970 to 1985 with an intermediate balance taken at 1980. These data are given in Tables 4 and 5.

It follows that, as a general rule, the NO_x emissions strongly increased during the years 1970-1980, and in the period from 1980 to 1985 tended to flatten out. An explanation of this phenomenen is not given by the author of the reference.
In the case of the emission of SO_2, the picture is different. Here, an overall decrease can be observed. The average reduction rates over the period from 1980 to 1985 is about 40%.

It is not at all clear how the predictions of the pollutant emissions from 1990 to 2000 will prove to be realistic. The present situation and future tendencies are largely uncertain.

	TOTAL SO$_2$ EMISSION in 1000 Metric Tonnes					Reduction Rate			Reduction Rate from 1980	
	Year	Amount	1980	Year	Amount	from	to	%	to	%
BELGIUM	1970	1 033	799	1984	467	1970	1984	55	1984	42
DENMARK	1970	523	438	1985	326	1970	1985	38	1985	26
FRANCE[1]	1971	2 966	3 558	1985	1 845	1971	1985	38	1985	48
GERMANY, F.R.[2]	1970	3 600	3 200	1985	2 400	1970	1985	33	1985	25
GREECE			800*	1983	720*				1983	10
IRELAND	1972	186	219	1985	138	1972	1985	26	1985	37
ITALY	1972	3 200	3 800	1985	2 503	1972	1985	22	1985	34
LUXEMBOURG	1970	35	23	1985	13	1970	1985	63	1985	44
NETHERLANDS[3]	1970	685	487	1985	275	1970	1985	60	1985	44
PORTUGAL	1970	116	266	1983	305	1970	1983	+ 163	1983	+ 15
SPAIN	1975	3 004	3 250	1983	3 250	1975	1983	+ 8	1983	0
UNITED KINGDOM[4]	1970	6 120	4 670	1985	3 580	1970	1985	42	1985	23

1) Total Emission 1987 (preliminary figures): 1,520 000 t/a
2) "   "   1987 "   "   : 2,000 000 t/a
3) "   "   1987 "   "   :   282 000 t/a
4) "   "   1987 "   "   : 3,680 000 t/a

* According to ECE from 1988, the total SO$_2$-emissions amounts to 400 kt in 1980 and to 360 kt in 1983. If these figures are correct Greece would have a better position in the ranking list.

Table 4: Development of the SO$_2$-emissions in Europe
Source: (11)

TOTAL NO$_x$ EMISSION
in 1000 Metric Tonnes

	Year	Amount	1980	Year	Amount	Change Rate from	to	%	Change Rate from	to	%
BELGIUM	1970		442	1984	385				1970	1984	- 13
DENMARK	1970	192	251	1985	238	1970	1985	+ 19	1970	1985	- 5
FRANCE[1]	1973	1 699	1 867	1985	1 600[a]	1973	1985	- 0,4	1973	1985	- 9
GERMANY, F.R.[2]	1970	2 400	3 100	1985	2 900	1970	1985	+ 17	1970	1985	- 7
GREECE			127	1983	150					1983	+ 15
IRELAND	1972	55	67	1985	68	1972	1985	+ 19	1972	1985	+ 1,5
ITALY			1 480[b]	1985	1 595					1985	+ 7,2
LUXEMBOURG			23	1985	22					1985	- 4
NETHERLANDS	1970	398	535	1985	537	1970	1985	+ 26	1970	1985	0
PORTUGAL	1970	72	166	1983	192	1970	1983	+ 63	1970	1985	+ 14
SPAIN	1975	624	ca. 950	1985	950[c]	1975	1985	+ 34	1975	1985	0
UNITED KINGDOM	1970	2 033	1 916	1985	1 837	1970	1985	- 10	1970	1985	- 4

1) Total NO$_x$ emission 1987 (preliminary figures): 2,350 000 t/a
2) "   "   "   1986 "   " : 3,000 000 t/a
3) "   "   "   1987 "   " : 1,990 000 t/a

a) Excluding emissions by agricultural activities estimated to be about 700 kt/a
b) Arithmetic mean; basic data: 1 410 - 1 550 000 Metric Tonnes
c) Arithmetic mean; basic data: 779 - 1 220 000 Metric Tonnes

Table 5: Development of the NO$_x$-emissions in Europe
Source: (11)

4. References

[1] Environmental resources Limited for CEC (1983) Acid Rain, Graham and Trotman Limited, London

[2] Weider, H. (1986). Air pollution control in the FRG; laws, regulation, implementation and principal shortcomings, Sigma Bohn, Berlin

[3] Schäfer, H.- G. (1987), conference papers:
Die Bildung von Emissionen während des Kraftwerkbetriebes, ihre Folgen und Verhinderung, Essen

[4] VGB-Handbuch (1986) NO_x-Bildung und NO_x-Minderung bei Dampferzeugern für fossile Brennstoffe: Teil BI, NOx-Bildung bei der Verbrennung, Essen

[5] OECD Environmetal Data, Compendium 1989, Paris

[6] Goethel, G.F. (1982) Entstehung und Verhütung von Emissionen in der Mineralölindustrie, Staub-Reinhaltung der Luft 42, 465 - 469

[7] Winkler, K. (1982) Entstehung und Verhütung von Emissionen in der chemischen Industrie, Staub-Reinhaltung der Luft 42, 459 - 464

[8] Gerold, F. (1980) Emissionsfaktoren für Luftverunreinigungen, Umweltbundesamt Materialien 2/1980, Erich Schmidt, Berlin

[9] Umweltbundesamt (1989) Daten zur Umwelt 1988/89, Erich Schmidt, Berlin

[10] Kuhler, M. (1986) Natürliche und antropogene Emissionen, GWF-Gas/Erdgas 127 H./, 27 - 35

[11] Weidner, H. (1987). A survey of clean air policy in Europe, Wissenschaftszentrum Berlin für Sozialforschung GmbH

ATMOSPHERIC CHEMISTRY OF SULPHUR DIOXIDE AND NITROGEN OXIDES

G. Restelli and H. Stangl
Commission of the European Communities
Joint Research Centre - Ispra Site
Environment Institute - Air Chemistry
I-21020 ISPRA (VA), Italy

1. Introduction

Burning fossil fuels is responsible for injecting in the atmosphere, in addition to water and carbon dioxide, gases in variety and amount dependent upon the caracteristics of the fuel and of the combustion process.

The emission of sulphur dioxide and nitrogen oxides have been subject of special attention from the time they have been recognized responsible for local and regional pollution episodes, eventually producing significant damage to man and environment.

Understanding of the global environment is now putting anthropogenic pollution by combustion processes in a more extended perspective for the contribution to environmental issues not only of reactive pollutants via their atmospheric chemistry but also for the consequences of the emission of the less reactive (CO, CH_4) and inert (CO_2, N_2O) gases. The increasing levels of trace gases in the atmosphere are infact considered potentially responsible for climatic modifications at the planetary scale resulting either by direct perturbations of the radiative budget or by the influence, in combination with reactive species, on other radiatively important atmospheric components like O_3, aerosols and clouds (Wuebbles et al., 1989).

A review of the chemistry of the atmosphere is well beyond the scope of this paper; to this end the reader is referred to books with an exhaustive treatment of this topic (e.g. Seinfeld, 1985; Wayne, 1985; Finlayson-Pitts and Pitts, 1986). Here the tropospheric chemistry of SO_2 and NOx ($NO + NO_2$) will be reviewed in some detail with limited extension to that of other species, organics in particular, which is essential to a complete understanding of the role and fate of NOx and SO_2.

Both sulphur dioxide and nitrogen oxides emissions end up with the formation of sulphuric and nitric acid or of their salts. The intermediate fate however of these two pollutants and their impact on the chemistry of other trace gases are quite different; for this reason they will be treated separately. Interconnections between the chemistry of these two gases and their fate e.g. through the modifications induced by NOx on the oxidizing capacity of the atmosphere in turn affecting the oxidation of SO_2 will be evident from the discussion.

From an historical point of view, sulphur dioxide is bound to the episode of the

London smog in 1950s, later to the acidification of lakes, to visibility degradation and to the atmospheric corrosion of materials. SO_2 was finally involved together with NOx, in the still discussed forest die-back issue.

Nitrogen oxides have been identified as a key precursor, together with hydrocarbons and sunlight, of photochemical oxidants responsible for the Los Angeles sindrome; a phenomenon now recognized affecting many other cities in the world characterized by intense traffic and by the occurrence of particular meteorological conditions (strong solar irradiation and weak local wind circulation).

SO_2 and NOx exhibit a high reactivity and undergo fast chemical transformations to cause environmental damage on the local and regional scale; long range transport of acidic aerosols generated from the chemical transformation of these pollutants, has also been observed. The effects have been demonstrated to be sufficiently dramatic to justify the implementation of costly abatement technologies.

The evaluation of their impact on a global scale is still a matter of discussion. SO_2 and NOx are not involved in direct radiative forcing effects since their absorptions in the short-wavelength and in the long-wavelength part of the Earth's radiation spectrum are not relevant. An influence however of SO_2 and NOx on other components of the planetary radiation budget like tropospheric ozone and atmospheric albedo, via their atmospheric chemistry, appears to be potentially important. This point will be then discussed together with some aspects concerning N_2O, another non reactive oxide of nitrogen emitted in the combustion.

2. Atmospheric Chemistry: First Principles

The fate of a gas injected in the atmosphere is determined by its physico-chemical properties and by the characteristics of the background air.

Few species emitted as primary pollutants have high solubilities to be directly removed by dissolution in liquid water droplets in the troposphere (clouds, fog) and then deposited to the ground (wet deposition).

If the gas is unreactive it will spread through the troposphere; subsequently it may be transported to the stratosphere where the energetic actinic flux will decompose it, mostly into reactive species.

If the gas is not chemically inert, which is the case for most atmospheric species injected in reduced form from natural and anthropogenic sources, it will be oxidized. Oxidized species generally exhibit a higher water solubility and are substantially removed by dissolution and wash-out to the ground. Oxidized products are also frequently characterized by lower vapour pressures and may act as nucleation centers to form particles.

Primary and secondary pollutant gases can be finally directly deposited to the ground without having been firstly dissolved, in a process termed "dry deposition", recognized of increasing importance in the balance of many pollutants.

In conclusion, trace gases injected into the atmosphere will disappear by chemical reactions, by dissolution followed by deposition or by direct deposition to the ground; the balance between the strength of the sources and the rate of loss by these processes determines the steady state atmospheric concentration.

The presence of molecular oxygen as one of the two main components (N_2:79%; O_2:21%) of the air appears to give it the obvious role of the oxidizing species; this is only true from the point of view of the ultimate source of oxygen. The kinetics of the process is in fact determined by reactive species, present at concentrations many orders of magnitude smaller than molecular oxygen, but characterized by substantially higher reactivities. These species are radicals generated in the troposphere by the simultaneous occurrence of two circumstances: the existence at the Earth's surface of a solar flux extending towards the short-wavelength part of the spectrum to about 300 nm and the presence of molecules capable to dissociate under absorption of this part of the solar radiation spectrum.

Most gas-phase oxidation processes are either directly or indirectly initiated as a result of UV solar absorption.

In this context a pivotal role is taken by the hydroxyl radical, OH. This highly reactive photolytic species, present in the atmosphere at daytime concentrations of $10E(5)$-$10E(7)$ radicals cm^{-3} (i.e. about $10E(-13)$ times that of O_2) appears to control the chemical lifetime of nearly all the atmospheric trace gases.

A role of importance comparable to that of the OH radical during daytime is played by the nitrate radical, NO_3. This species, generated by the reaction between NOx and O_3 and rapidly photolyzed in daylight, has been shown in these last years to determine the chemistry of the perturbed troposphere at night (Wayne et al., 1991).

Other radicals, HO_2, RO, RO_2 (R=alkyl group) etc... are less involved in the initial oxidative attack of atmospheric trace gases, but are important in the overall process.

Only a limited number of the compounds present in the atmosphere are capable to undergo photolytic dissociation reactions leading to radical generation, under absorption of the UV solar spectrum at ground level; O_3, NO_2, HONO, H_2O_2, HCHO are by far the most important.

Ozone, O_3, is the primary source of OH radicals both in the clean and in the polluted troposphere. Following absorption in the Hartley, Huggins and Chappuis bands, O_3 undergoes dissociation as:

$$O_3 + h\nu \longrightarrow O_2 + O(3P) \qquad h\nu < 610 \text{ nm} \qquad (1)$$

$$O_3 + h\nu \longrightarrow O_2 + O(1D) \qquad h\nu < 310 \text{ nm} \qquad (2)$$

Reaction (2) is followed by quenching of the excited O(1D) atom by collision with nitrogen and oxygen molecules to O(3P), which rapidly recombines with O_2 to regenerate O_3:

$$O(3P) + O_2 + M \longrightarrow O_3 + M \qquad (3)$$

O(1D) has a fairly long radiative lifetime (110 s) and in alternative to colliding with N_2 or O_2 it can react with H_2O vapour molecules as:

$$O(1D) + H_2O \longrightarrow 2 \text{ OH} \qquad (4)$$

to produce hydroxyl radicals. The rate constant of reaction (4) is faster than that of the collisional deexcitation. About 1% of the excited O atoms may then be transformed into OH radicals; with equatorial sun, 30 ppbv of O_3 and 20 torr of H_2O this represents a source of $10(+7)$ OH radicals $cm^{-3}s^{-1}$.

The O_3 photolysis process at $h\nu > 310$ nm is not significant for the tropospheric chemistry.

Nitrous acid, HONO, according to reaction (5):

$$HONO + h\nu \longrightarrow OH + NO \qquad h\nu < 400 \text{ nm} \qquad (5)$$

and H_2O_2 according to reaction (6):

$$H_2O_2 + h\nu \longrightarrow 2OH \qquad h\nu < 360 \text{ nm} \qquad (6)$$

are also sources of OH radicals.
The first one is of importance in the polluted troposphere, (see paragraph 4.1).

H_2O_2, easily dissolved in water, is itself an important oxidant and a radical source for oxidation processes taking place in the liquid phase.

Nitrogen dioxide, NO_2, and formaldehyde, HCHO, are the other important precursors of photolytic radicals. These gases are respectively a source of oxygen atoms O(3P) and of HO_2 radicals according to the reactions:

$$NO_2 + h\nu \longrightarrow NO + O(3P) \qquad h\nu < 430 \text{ nm} \qquad (7)$$

$$HCHO + h\nu \longrightarrow HCO + H \qquad h\nu < 330 \text{ nm} \qquad (8)$$

followed respectively by reaction (3) and by reactions (9) and (10):

$$H + O_2 + M \longrightarrow HO_2 + M \qquad (9)$$

$$HCO + O_2 \longrightarrow CO + HO_2 \qquad (10)$$

NO from reaction (7) in turn can react with the HO_2 radical in the very important reaction (11):

$$NO + HO_2 \longrightarrow OH + NO_2 \qquad (11)$$

to generate OH radical. In the atmospheric degradation of organics and other trace gases this reaction is of central importance for the regeneration of the OH radicals and the propagation of oxidative chain reactions.

In presence of O_3, NO is oxidized to NO_2 by the relatively fast reaction:

$$NO + O_3 \longrightarrow NO_2 + O_2 \qquad (12)$$

In the case however that NO is oxidized to NO_2 by another reaction (e.g. reaction (11)), the system of reactions (7,3), capable to mediate between the energy needed to photolyze an oxygen molecule (see below) and the photon energy limit of the tropospheric solar spectrum, is a net source of ozone.

In the troposphere, ozone of this chemical origin is indicated as "photochemical" in contrast to the one that is injected by downward transport from the stratosphere, where it is formed by dissociation of molecular oxygen (hv<240 nm). Contributions of these two sources may be comparable on a global scale; the photochemical source may even largely exceed the other one in polluted areas.

The other fundamental component of the reacting mixture in the atmosphere is represented by the whole of reactive compounds of the biogeochemical cycle of carbon. Incomplete combustion of fossil fuels, industry and natural emissions (methane, isoprene and monoterpenes from vegetation), are the sources of these species. Apart from CO_2, unreactive in the lower atmosphere, CO and CH_4 are the most abundant gases of the carbon cycle; other organic compounds, usually indicated by the abbreviation "VOC", volatile organic compounds, while injected in a total quantity comparable to that of CH_4, are present at much reduced concentrations due to their higher reactivities and shorter atmospheric lifetimes.

In the remote troposphere OH radicals react with carbon monoxide and methane according to known chemical reactions which can be used as representative of most atmospheric oxidation schemes.

For the oxidation of carbon monoxide, CO:

$$CO + OH \longrightarrow CO_2 + H \tag{13}$$

$$H + O_2 + M \longrightarrow HO_2 + M \tag{9}$$

$$HO_2 + NO \longrightarrow OH + NO_2 \tag{11}$$

$$NO_2 + h\nu \longrightarrow NO + O(3P) \tag{6}$$

$$O(3P) + O_2 + M \longrightarrow O_3 + M \tag{3}$$

and for methane, CH_4:

$$CH_4 + OH \longrightarrow CH_3 + H_2O \tag{14}$$

$$CH_3 + O_2 + M \longrightarrow CH_3O_2 + M \tag{15}$$

$$CH_3O_2 + NO \longrightarrow CH_3O + NO_2 \tag{16}$$

$$CH_3O + O_2 \longrightarrow HCHO + HO_2 \tag{17}$$

the formaldehyde following photolysis or attack by OH radicals, also leads to formation of HO_2 radicals, rapidly recycling OH in the presence of NO and leaving as a net balance a build-up of O_3. It is evident that molecular oxygen is consumed and the hydroxyl radical acts as catalist in the oxidation chain. The chain can however be terminated by radical radical reactions such as:

$$CH_3O_2 + HO_2 \longrightarrow CH_3OOH + O_2 \tag{18}$$

$$NO_2 + OH + M \longrightarrow HNO_3 + M \tag{19}$$

$$HO_2 + HO_2 \longrightarrow H_2O_2 + O_2 \tag{20}$$

CH_3OOH and HNO_3 whose photolysis is slow, can be rained out or deposited at the earth's or aerosols' surface; H_2O_2, in the gas phase or dissolved in water, can be subsequently photodissociated, thus serving as a temporary reservoir of OH radicals.

Alkanes, alkenes, aromatics and other volatile, natural and anthropogenic, organic pollutants react according to analogous but more complex and extended reaction sequences, initiated by H-atom abstraction or OH addition to unsaturated bonds, alkyl radical reaction with O_2, isomerization and decomposition, to produce intermediates and end products like carbonyl species, organic nitrates, etc...

The role of the OH radical was recognized at the beginning of the 1960s (Leighton, 1961); in the following years the oxidative reaction chains of hydrocarbons and CO were proposed (Levy, 1971) and the hypotheses experimentally confirmed. It was however only in the 1970s that the importance of the nitrate radical NO_3 in the nighttime chemistry emerged (Morris and Niki, 1974), especially after its detection in clean (Noxon et al., 1978) and in polluted (Platt et al., 1980) air at concentrations ranging from $10(+8)$ to $10(+10)$ radicals cm^{-3}. The nitrate radical in fact reacts with some trace gases at rates sufficiently high to become a sink comparable to that represented in daylight by the reaction of these species with OH radicals.

The NO_3 radical which is rapidly photolized in daylight, is formed at night by the reaction:

$$NO_2 + O_3 \longrightarrow NO_3 + O_2 \tag{21}$$

and establishes the fast equilibrium:

$$NO_3 + NO_2 + M \longleftrightarrow N_2O_5 + M \tag{22}$$

which limits the concentration of NO_3 in the atmosphere. N_2O_5 acts as a reservoir for the NO_3 radicals.

Like in the case of OH, oxidation by NO_3 attack is initiated by H-atom abstraction or by addition to double bonds.

An important difference with respect to OH is, in this case, the formation of HNO_3 or of organic nitrates as reaction products, subsequently scavenged to the surface by wet and dry deposition. Organic nitrates are of particular environmental concern since they have been recognized noxious to vegetation and suspected for mutagenicity. Organic nitrates can also be formed in the OH attack of organics in an NO_2 rich atmosphere; chemistry and physics of these compounds have been recently exhaustively reviewed (Atkinson, 1990; Roberts, 1990).

The discussion has up to now concerned the homogeneous gas phase chemistry; in these last years an increasing attention is being dedicated to reactions that involve atmospheric gases and occur in the liquid water phase and on the surface of solid aerosols. Heterogeneous reactions are however much more difficult to study than homogeneous gas reactions and many questions are still open.

Heterogeneous reactions have been recognized to substantially contribute to SO_2 oxidation and are now under extensive discussion for the role the clouds may play in the chemistry of the natural troposphere (Lelieveld and Crutzen, 1990). H_2O_2 and ozone, together with their photolysis products, appear to be the main oxidizing species in the liquid phase; both are however originated in the gas phase which then appears to control also the atmospheric chemistry taking place in the atmospheric liquid phase.

It should be evident by now that, in the gas phase, NOx emissions may affect the concentration of O_3 and OH, significantly influencing the oxidizing capacity of the troposphere. NOx are also necessary for the formation of NO_3 radicals and then responsible for the atmospheric chemistry developing at night.

A role of SO_2 in this framework is not evident and emissions of this gas should have a minor impact on the overall chemistry of the troposphere in spite of their evident adverse effects on the environment.

Contrary to what happens in the gas phase, SO_2 is likely to affect the liquid phase chemistry, since an important fraction of the oxidation of SO_2 to sulphate occurs in this state. A role of NOx in the liquid phase processes may occur only indirectly through the influence exerted over the formation of gas phase oxidants.

3. The Atmospheric Chemistry of SO_2

As anticipated, the atmospheric chemistry of SO_2 is much simpler than that of the nitrogen oxides. SO_2 is not photosensitive to sunlight radiations in the troposphere despite considerable absorption in the near UV; the absorbed energy is insufficient to break the OS-O bond (threshold at hv=218 nm) and the excited states are rapidly quenched without chemical effects.

The oxidation of SO_2 to sulphuric acid and sulphates is rather straightforward and sideproducts are almost absent. Several mechanisms are however active for this transformation, both in the gas and in the liquid phase.

3.1 HOMOGENEOUS GAS PHASE REACTIONS

Sulphur dioxide has a strong tendency to react with O_2 to form SO_3: at 298 K and 1 atm in air the equilibrium concentration ratio, $[SO_3]/[SO_2]$, is equal to $\sim 10\ E(12)$. Yet, in absence of catalysts, the rate of the reaction in the gas phase is so slow that oxidation of SO_2 by O_2 can be completely neglected. Similarly negligible is the photooxidation of SO_2 (Sidebottom et al., 1972).

The reaction of SO_2 with ozone is also too slow to be significant; a rate constant for this reaction at 1 atm and 298 K equal to or smaller than $8\ 10E(-24)$ $cm^{+3} molecule^{-1} s^{-1}$ (Calvert and Stockwell, 1984), has been evaluated.

It is now clear that in the gas phase, the only significant chemical loss of SO_2 occurs via the reaction with the OH radicals as:

$SO_2 + OH + M \longrightarrow HOSO_2 + M$ (23)

with an effective bimolecular rate constant equal to $9\ E(-13)$ $cm^{+3} molecule^{-1} s^{-1}$ at 1 atm and 298 K (Calvert and Stockwell, 1984).

Oxidative reactions with other species, O(3P), HO_2, are estimated to be much less important. Rates and products of the reactions of $HOSO_2$ in the polluted troposphere are not fully understood. The reaction with molecular oxygen should be the predominant fate; addition (Davis et al., 1979):

$$HOSO_2 + O_2 + M \longrightarrow HOSO_2O_2 + M \qquad (24)$$

and H-atom abstraction (Stockwell and Calvert, 1983)

$$HOSO_2 + O_2 \longrightarrow HO_2 + SO_3 \qquad (25)$$

have been proposed.

In support to this last mechanism which is presently the most accredited, is the rate constant for reaction (25) measured as fast as (4 ± 2) E(-13) $cm^{+3}molecule^{-1}s^{-1}$ at 298 K (Margitan, 1984). Reaction (23) appears to be the rate limiting step in the process.

Reaction (25) forms HO_2 radicals: this is important because HO_2 can react with NO to regenerate OH or can undergo the selfreaction (20) forming H_2O_2, which is a primary oxidant of SO_2 in the liquid phase.

The SO_3 rapidly reacts with H_2O to form sulphuric acid:

$$SO_3 + H_2O \longrightarrow H_2SO_4 \qquad (26)$$

The only other important homogeneous gas phase reaction of any importance for the oxidative removal of SO_2 is dependent on the simultaneous presence of O_3 and of unsaturated hydrocarbons (Cox and Penkett, 1971, 1972). With O_3 as an ubiquitous pollutant, this process would depend on local olefin emissions. A special case with an unknown general impact is represented by the biogenic emissions of unsaturated hydrocarbons (monoterpenes, isoprene) from vegetation.

The oxidation reaction of SO_2 must be attributed to intermediates, produced in the O_3 olefin reaction, in particular to the Criegee biradical (Martinez et al., 1981; Martinez and Herron, 1981), e.g.:

$$SO_2 + CH_3CHOO \longrightarrow SO_3 + CH_3CHO \qquad (27)$$

Estimates of the rate constant for reaction (27) range from 2 E(-11) to 5 E(-15) $cm^{+3}molecule^{-1}s^{-1}$ (Herron et al., 1982). Reaction (27) can occur at night when also the other important atmospheric radical, NO_3, is ineffectual for the oxidation of SO_2; in fact, an upper limit for the rate constant between NO_3 and SO_2 equal to ~10 E(-21) $cm^{+3}molecule^{-1}s^{-1}$ has been estimated (Daubendiek and Calvert, 1975).

In the case of SO_2-O_3-terpene reaction no sulphur containing organics were found, which demonstrates that in contrast to the corresponding reaction of O_3, NO_2 and olefins, which often leads to organic nitrates, no sulphonization takes place (Kotzias et al., 1990). Under most atmospheric conditions reaction (27) cannot be an important route for SO_2 oxidation; the question whether it may have a role in special situations is open (Stangl et al., 1986).

3.2 HETEROGENEOUS REACTIONS

SO_2 can also be oxidized when absorbed on aerosol particles, or dissolved in liquid water, cloud or fog droplets (homogeneous liquid phase). In early studies of SO_2 to sulphate conversion a catalytic action of metal ions like Fe^{2+} and Mn^{2+} in fly-ash was considered to promote the oxidation by molecular oxygen of the absorbed SO_2. In the following years, the process of transfer of SO_2 from the gas phase to the liquid phase and its oxidation in the liquid phase have been subject of extended research. Several oxidizing species have been identified, including dissolved oxygen, ozone and hydrogen peroxide, with the reactions also catalyzed by dissolved metal ions.

The most important reaction of SO_2 in the atmospheric liquid water is believed to be that with H_2O_2, and perhaps that with O_3 at high pH values. SO_2 is soluble in water (Henry's law coefficient equal to 1.24 mole L^{-1} atm^{-1} at 298 K) and the rate determining step appears to be the chemical reaction. It is not clear whether the reaction with H_2O_2 involves all the S(IV) species or only the bisulphite ion in the generally accepted mechanism:

$$HSO_3^- + H_2O_2 \longrightarrow SO_2OOH^- + H_2O \tag{28}$$

$$SO_2OOH^- + H^+ \longrightarrow H_2SO_4 \tag{29}$$

The rate of oxidation appears independent of the pH value and only below pH=1 the rate decreases as the pH decreases.

The O_3-S(IV) reaction proceeds via three independent paths (Hoffmann, 1985) involving a nucleophilic attack on O_3 by $SO_2 \cdot H_2O$, HSO_3^- and SO_3^{--} as e.g.

$$HSO_3^- + O_3 + OH^- \longrightarrow SO_4^{--} + H_2O + O_2 \tag{30}$$

The rate of the reaction with ozone (50 ppbv in the gas phase) increases substantially with pH to exceed that of the reaction with H_2O_2 (1 ppbv in the gas phase) at pH values over 5.5 (Martin, 1984).

The oxidation by dissolved O_2 catalyzed by Fe, the most abundant atmospheric transition metal, or by Mn has been investigated by several groups; the oxidation rates are comparable to those of the reactions above discussed, only at high pH values (see in Hoffmann and Calvert, 1985).

Oxidation by NO_2 or by HNO_2 (NO_2^-) appear unimportant at all pH values (Martin et al., 1981; Lee and Schwartz, 1982).

Oxidation rates of SO_2 by H_2O_2(aq) up to 100% per hour have been observed; for the same conditions the rates for the oxidation by O_2 catalyzed by Fe or Mn are below 1% per hour at pH values in the solution less than 4.5.

The reaction rates for the SO_2 oxidation in the liquid phase increase at decreasing temperatures, due to the predominant effect of the increased SO_2 solubility.

The photochemical activity in the gas phase promotes the oxidation of SO_2 both in the gas and in the liquid phase and limits, with the supply of oxidants, the effective rate of transformation of SO_2 to sulphate. H_2O_2 can be absorbed from the gas phase, where it is generated by reaction (20), or created in the solution by dark and by photochemical reactions; O_3 is dissolved in water with a Henry's law coefficient at 298 K equal to 0.01 mole L^{-1} atm^{-1}.

A certain role of reaction inhibitor has been attributed to formaldehyde, a product of the photooxidation of hydrocarbons in air, easily soluble in water droplets. Formaldehyde creates hydroxymethylsulphonate ions:

$$HOSO_2 + HCHO \longrightarrow HOCH_2SO_3^- \tag{31}$$

which are oxidized at slower rates. For an exhaustive analysis of the large literature existing on the reaction kinetics of aqueous sulphur chemistry, the reader is referred in particular to Ref. (Calvert, 1984).

It seems opportune to emphasize the total dependence of the SO_2 oxidation processes on photolytic radicals generated in the gas phase reactions: OH and HO_2 radicals control the oxidizing capacity in the gas and in the liquid phase. This explains the lower sulphate concentration in precipitations observed in winter with respect to those measured in summer; a situation which does not occur in the case of nitrates which show little seasonal variation for the contribution of the nighttime NOx to HNO_3 conversion via the nitrate radical NO_3.

The end product of the gas phase oxidation of SO_2 is then a fine, but rapidly coagulating sulphuric acid aerosol which grows into larger sized particles. The high aerosol concentration is responsible for the appearance of a dense haze, especially at high relative humidities. Normally only a fraction of the oxidized SO_2 is found through the atmosphere as free sulphuric acid. It is estimated that over continental areas the presence of free sulphuric acid would be no more than 20% of the total sulphate ions, due to the presence of neutralizing species, mainly ammonia. Wet deposition depends mainly upon the amount of rainfall. The dry deposition of SO_2 has been found important in the atmospheric balance of this gas; this process has been estimated to remove SO_2 with an efficiency comparable to that of the chemical transformations followed by wet deposition.

The relative importance of wet and dry deposition as sulphur removal processes from the atmosphere in the case of SO_2 emissions have been discussed by several authors for the cases of unpolluted and of polluted areas; the dry deposition has been found to account for a fraction varying from 1/2 to 2/3 of the total.

4. Nitrogen Oxides

4.1 NITRIC OXIDE

Burning any combustible material in air leads to generation of NO in a complex scheme of radical chain reactions, favouring at high temperatures the formation of this gas.

In air, nitric oxide is however liable to further oxidation. Oxidation by molecular oxygen is not significant. The oxidation in fact follows a termolecular reaction:

$$2NO + O_2 \longrightarrow 2NO_2 \tag{32}$$

with a rate constant at 298 K equal to 2.0 10E(-38) cm^{+6}molecule^{-2}s^{-1} (Hampson and Garvin, 1978), so that only at relatively high concentrations of NO as in stacks, the process may occur at a reasonable rate. At the typical low levels found in the troposphere oxidation of NO to NO_2 by this reaction would require days.

When diluted in ambient air, NO is mainly oxidized by the reaction with O_3

$$NO + O_3 \longrightarrow NO_2 + O_2 \tag{12}$$

with a rate constant at 298 K equal to $1.8 \cdot 10E(-14)$ cm^{+3}molecule^{-1}s^{-1} and by the reaction with peroxyradicals RO_2 (R = H, CH_3 or another alkyl group):

$$NO + RO_2 \longrightarrow RO + NO_2 \tag{33}$$

with a rate constant of the order of 8 10E(-12) cm^{+3}molecule^{-1}s^{-1} at 298 K (DeMore et al., 1990).

In polluted air when direct sources of NO are present, the formation of O_3 is inhibited until the NO concentration has decreased below a certain value.

As pointed out in paragraph 2, the fast reactions (11,33) are central in clean and in polluted air to any photochemical process involving chain reactions characterizing the oxidative degradation of organics initiated by the attack of OH radicals.

Other reactions involving NO in ambient air are the reaction with OH (RO) radicals:

$$NO + OH (RO) + M \longrightarrow HONO (RONO) + M \tag{34}$$

resulting in the formation of nitrous acid or of organic nitrites, and the fast reaction with the nitrate radical:

$$NO + NO_3 \longrightarrow NO_2 + NO_2 \tag{35}$$

Reaction (34) is in the falloff region between second and third order at 1 atm and 298 K with a rate constant (2nd order) equal to $6.8 \cdot 10E(-12)$ cm^{+3}molecule^{-1}s^{-1} (Atkinson and Lloyd, 1984). The transformation of NO into HONO may also have heterogeneous paths and the following reactions have been suggested:

$$NO + OH (aq) \longrightarrow HONO (aq) \tag{36}$$

$$NO + NO_2 + H_2O(l) \longrightarrow 2 HONO (aq) \tag{37}$$

with rate constants respectively equal to $1 \cdot 10E(10)$ Lmole^{-1}s^{-1} and $3 \cdot 10E(7)$ Lmole^{-1}s^{-1} (Graedel and Goldberg, 1983). In polluted urban air HONO is a major source of OH radicals at dawn, when it is rapidly photolyzed according to reaction (4).

Reaction (35) is a sink for the NO_3 radical; its rate constant is equal to 2.9 10E(-11) cm^{+3}molecule^{-1}s^{-1} at 298 K (DeMore et al., 1990).

The lifetime of NO, γ , (in the case of bimolecular reactions, $\gamma = 1/k[X]$ where k is the rate constant of NO with the species X of concentration [X]) in polluted air, is of the order of tens of seconds for the RO_2, O_3 and NO_3 reactions, of hours for the OH reaction. It may be noteworthy the fact that NO has no chemical sinks in the troposphere, since only tempory reservoirs are formed. This is the case of NO_2, HONO etc... which regenerate NO following photolysis.

4.2 NITROGEN DIOXIDE

4.2.1 Homogeneous Gas-Phase Processes. Some percent of the nitrogen oxides emitted from combustion processes may occur as NO_2 so that this species is both a primary and a secondary pollutant. As anticipated, the role of this species in the atmosphere is central to any oxidation process leading to transformation of other trace gases injected in the atmosphere by natural or anthropogenic sources. According to reaction (7) NO_2 is photodissociated into NO and an oxygen atom in the ground state O(3P). The quantum yield close to zero at 430 nm, rises rapidly to 0.8 at the theoretical threshold of 397.8 nm and then slowly increases to reach 1 at 300 nm. The non zero yield above the threshold wavelength is attributed to the contribution of internal rotational energy and the yield less than one, at wavelengths below 397.8 nm, to the formation of a non dissociative state. The lifetime for photolysis of NO_2 in the atmosphere is obviously dependent on latitude, season, solar zenith angle, sky conditions; in general it is of the order of a few minutes.

The O(3P) atoms react with O_2 with a third order rate constant ($6 \cdot 10E(-34)$) $cm^{+6} molecule^{-2} s^{-1}$) which at 1 atm leads to a lifetime for the oxygen atoms as short as 13 μs.

In the ideal case that no other reactive species are present, O_3 rapidly reacts with NO as in reaction (12) to reform NO_2 and O_2. Reactions (7), (3) and (12) establish an equilibrium which holds as long as reaction (12) is the principal oxidation process for NO and the only loss for O_3. The concentrations of O_3, NO and NO_2 are linked by the Leighton relationship:

$$[O_3][NO]/[NO_2] = J7/k12 \tag{38}$$

where J7 and k12 are the rate constants for reactions (7) and (12). J7, dependent on solar zenith angle, season, latitude, etc.. is $< 9 \cdot 10E(-3) s^{-1}$ and k12 is at 298 K equal to $1.8 \cdot 10E(-14)$ cm^{+3} $molecule^{-1} s^{-1}$.

This relation shows that the photolysis of NO_2 leads to small, but non zero, steady state concentrations of O_3 and of NO. NO is present also in excess O_3 and this is important for the role it plays for the recycling of OH via HO_2 reduction (reaction (11)).

At night when NO (NO_2) and O_3 are simultaneously present, NO_2 reacts with O_3 (reaction (21)) with a rate constant equal to $3.2 \cdot 10E(-17)$ $cm^{+3} molecule^{-1} s^{-1}$ at 298 K (DeMore et al., 1990). NO_2 reacts further with NO_3 and establishes the fast equilibrium 22: Keq ~$4 \cdot 10E(-11)$ $cm^{+3} molecule^{-1} s^{-1}$ at 298 K (see discussion on the value of Keq and its importance in Finlayson-Pitts and Pitts, 1985; Wayne et al., 1991). The equilibrium is strongly dependent on temperature (displaced to the right at lower temperatures) so that N_2O_5 may act as an efficient reservoir of the NO_3 radical.

NO_2 reacts with NO_3 also as:

$$NO_2 + NO_3 \longrightarrow NO + NO_2 + O_2 \tag{39}$$

but this channel is orders of magnitude slower (Graham and Johnston, 1978; Hjorth et al.,1989).

NO_3 is rapidly photolized in daylight with regeneration of NO_2:

$$NO_3 + h\nu \longrightarrow NO_2 + O \qquad h\nu < 670 \text{ nm} \qquad (40)$$

$$\longrightarrow NO + O_2 \qquad (41)$$

with channel 40 predominant over channel 41 (Magnotta and Johnston, 1980).

Due to a reactivity of NO_3 versus many organic species comparable to that of OH, and to the reaction of N_2O_5 with H_2O (see later in this paragraph), NO_3 represents at night, an efficient pathway to NOx removal and a source of radicals.

The reactions of NO_2 with other species, which occur in the perturbed atmosphere, can be divided into two categories: those leading to temporary reservoirs and those permanently removing NO_2.

The reactions with peroxydes belong to the first category:

$$NO_2 + HO_2 + M \longleftrightarrow HO_2NO_2 + M \qquad (42)$$

$$NO_2 + RO_2 + M \longleftrightarrow RO_2NO_2 + M \qquad (43)$$

All these compounds are thermally unstable and decompose back to the reactants.

The most important representative of this category is the peroxyacetylnitrate (PAN). PAN is formed in the reaction of NO_2 with the acetylperoxyradical, an intermediate of the acetaldehyde attack by OH radicals, in the equilibrium:

$$CH_3COO_2 + NO_2 + M \longleftrightarrow CH_3COO_2NO_2 + M \qquad (44)$$

displaced to the right at low temperatures. PAN, an indicator of photochemical activity identified in any atmosphere from clean to urban (Singh and Hanst, 1981; Temple and Taylor, 1983) acts, when decomposed, as a non photolytic radical source. Its importance is mainly bound to the fact that this species, which is rather stable at the low temperatures found above a few kilometer altitude in the troposphere, may promote long range transport of NOx (Crutzen, 1979). In heavily polluted air, PAN has been found to account for 30-80% of the NOx oxidation products (Grosjean, 1983).

The reaction of NO_2 with the OH radical (reaction (19)), and the transformation of NO_2 into NO_3 at night (reaction (21)), belong to the second category.

The reaction of NO_2 with OH is fast, with a rate constant equal to $\sim 10E(-11)$ $cm^{+3} molecule^{-1} s^{-1}$; it represents the most important daytime sink for NO_2 and a chain reaction terminator for both NO_2 and OH radicals.

At $[OH] = 3 \cdot 10(E+6)$ cm^{-3}, considered typical for a summer daytime average (rural site), the conversion rate of NO_2 to HNO_3 is about 4% per hour, averaged over the whole day. This corresponds to a lifetime of 1 day that must be compared to the 13 days calculated for the SO_2 oxidation by OH (reactions (24), (25)) under the same conditions.

At low OH concentration the other NOx sink process dependent upon the formation of NO_3 may be of comparable significance.

NO$_3$ can be directly transformed into HNO$_3$ by the fast reactions, initiated by H-atom abstraction:

$$NO_3 + HX \longrightarrow HNO_3 + X \tag{45}$$

that the nitrate radical undergoes with organic species present in the atmosphere; e.g. with aldehydes in continental air (Cantrell at al., 1985 - 1987; Hjorth et al., 1988), or with dimethylsulphide (DMS), in marine air (Burrows et al., 1986; Jensen et al., 1990).

Also the reaction of NO$_3$ with alkenes, initiated by NO$_3$ addition to the double bond, represents a potential sink for NOx. The formation of organic nitrates is in fact a sometimes favoured reaction path (Hoshino et al., 1978; Shepson et al., 1985; Barnes et al., 1990; Hjorth et al., 1990). These products, especially those formed in the reaction between NO$_3$ and biogenic alkenes, are characterized by a low vapour pressure and may be rapidly deposited to the surface.

More important is the hydrolysis of N$_2$O$_5$ (originated from the fast equilibrium 22):

$$N_2O_5 + H_2O \longrightarrow 2\ HNO_3 \tag{46}$$

a process which implies the conversion of two molecules of nitrogen oxides for each N$_2$O$_5$ reaction.

The relative humidity needed for an efficient conversion of NO$_3$/N$_2$O$_5$ into HNO$_3$ appears to be only 50%; in most conditions, the reaction of NO$_2$ with O$_3$ (reaction (21)) is then the rate limiting step of the process.

The homogeneous gas phase reaction between N$_2$O$_5$ and H$_2$O is characterized by a slow rate constant (k~1 · 10E(-21) cm^{+3}molecule^{-1}s^{-1} at 298 K, Morris and Niki, 1973; Tuazon et al., 1983; Hjorth et al., 1987); and more importance is attributed to the N$_2$O$_5$ reaction with liquid water catalyzed by surface absorption (wet surfaces, water droplets), which is expected to be much faster (Platt et al., 1984).

4.2.2 Heterogeneous Processes. The atmospheric liquid water system, clouds, fog, rain, plays a double role with respect to the nitrogen oxides and their oxidation products.

On one side, gases physically dissolve in water and may, in this way, be scavenged to the ground without further reactions: this is the case of HNO$_3$ but not of PAN which is only slightly soluble. HNO$_3$ however, while it is irreversibly scrubbed in dilute solutions, in the case of concentrated solutions which are formed after cloud or fog evaporation, may return to the gas phase, because of its high vapour pressure. This occurs also in the case of neutralization with NH$_3$ either as concentrated solution or as dry salt.

NO and NO$_2$ are relatively insoluble in water (the Henry's law coefficients are respectively equal to 1.9 · 10E(-3) mole L^{-1} atm^{-1} and to 1 · 10E(-2) mole L^{-1} atm^{-1} at 298 K) and, at the partial pressures typical of the troposphere, aqueous phase nitrite and nitrate forming reactions are considered of minor importance except under the most extreme conditions (heavily polluted air and heterogeneous catalysis).

The liquid phase disproportionation of NO$_2$ in the equilibrium:

$$2NO_2 + H_2O(aq) \longrightarrow 2H^+ + ONO^- + NO_3^- \tag{47}$$

is now known to be too slow in the atmosphere and also the oxidation of NO_2 by H_2O_2 has been shown to be insignificant. On the contrary the high solubility of N_2O_5 and of HONO gives these species a potential role in the oxidation of other species in the liquid phase. For a detailed treatment of the aqueous phase reactions of the nitrogen oxides, the reader is referred to Refs. (Schwartz, 1984; Chameides, 1986).

The most relevant heterogeneous reaction in the NOx to HNO_3 transformation path, is the previously discussed hydrolysis of N_2O_5. It has been estimated that in summer the nighttime conversion of NOx to nitrate is of the same order of magnitude of the daytime conversion; in winter it can exceed the daytime conversion by an order of magnitude (Liu et al., 1987) and compensates for most part of the seasonal asymmetry expected from the daytime conversion only.

During the emission in the stack and in the power plant plume immediately before dilution in air, NOx has been found to be converted, in a mix of homogeneous and heterogeneous processes photochemically driven, at a rate from 0.2 to 12% per hour, mostly into HNO_3 and PAN. In the so-called urban plumes, rates of conversion of NOx to HNO_3 and PAN from less than 5% up to 24% per hour have been reported (Chang et al., 1979; Spicer, 1982).

In the gas phase HNO_3 is not significantly photolyzed and also the reaction with OH radicals is too slow to be of importance. The atmospheric lifetime of HNO_3 appears however short (about 3 days) due to dissolution into atmospheric water and to dry deposition. With regard to this last process, neutralization of HNO_3 produces a significant change in the lifetime for deposition, which is important in long range transport, since the aerosol nitrate deposits ten times more slowly then the acid.
The importance of the dry deposition of NO_2, NO and PAN, which however appears less efficient than in the case of SO_2, is still a matter of investigation and debate.

5. SO_2 and NOx Anthropogenic Emissions: A Climate-Chemistry Feedback?

The most evident adverse effects on the biosphere resulting from increased injection in the atmosphere of pollutants have been up to now observed on the local or regional scale; photooxidants build-up, visibility degradation, acidic deposition and corrosion of materials are some examples. The contribution of SO_2 and NOx and in particular of that part of these gases emitted from fossil fuel combustion (stationary and mobile sources) is quite determinant. In fact over continental areas corresponding to industrialized countries, this source of SO_2 and NOx dominates over those of natural origin. Most of the effects are limited in the spatial distribution to a regional scale by the rather short atmospheric lifetimes of both primary pollutants and transformation products.

Elevated concentrations of nitrates, and sulphates have been generally found in precipitations associated with zones of high emission densities.

In these last years much concern has been raised on the possibility of a modification of the planetary atmosphere induced by the increasing pollution, possibly leading to modifications of the Earth's radiative budget and to important climatic changes. The question may then be posed on the potential impact of the anthropogenic emissions of SO_2 and NOx in this context.

The discussion deliberately neglects the role of other important gases, CO_2, CO, CH_4, N_2O, emitted in the combustion of fossil fuels. For them the evaluation of the radiative forcing is straightforward, but much less obvious is the relation between the increase of the source strength and the variation in the global temperature.

The evaluation of the effect of gases such as SO_2 and NOx is more intriguing. These gases cannot be considered as radiatively active for the limited absorption in the short-wavelength and in the long-wavelength spectrum (low steady state concentrations due to their short atmospheric lifetimes). There are however reasons and some experimental evidence that these species may produce effects on important components of the planetary radiative budget influencing, via their role in atmospheric chemistry, the concentration of radiatively active gases (e.g. O_3) and the atmospheric albedo (aerosols/clouds formation and pattern).

The results of recent studies (Dickerson at al., 1987) which show a transport by thunderstorms of air pollutants to the upper troposphere, where they have much longer residence times and can move to much greater distances, reinforce this view.

5.1 TROPOSPHERIC OZONE

The reactions of O_3, H_2O, CH_4, CO and NOx, together with meteorological parameters like e.g. the solar photons flux, control the photochemistry of the background troposphere. In this framework the oxidation cycles of CO, CH_4 and other biogenic hydrocarbons may also have a regulatory role in the ozone budget of the troposphere. In the oxidation process of these gases (see paregraph 2), when the NOx level in the atmosphere exceeds that of ozone by the ratio between the rates of reaction (48):

$$HO_2 + O_3 \longrightarrow OH + O_2 + O_2 \tag{48}$$

and reaction (11)

$$HO_2 + NO \longrightarrow NO_2 + OH \tag{11}$$

ozone production by reaction (11) followed by reactions (7) and (3) becomes more important than ozone destruction by reaction (48).

This situation is reached in the troposphere for an NO mixing ratio of 5 pptv, equivalent to 15-20 pptv NOx. Taking into account also the ozone destruction by reactions (2) and (4), slightly higher NOx levels are required, depending on latitude and season. It is then clear that changes in the NOx background concentration, at this low level, might determine future trends in O_3. On the other hand, the O_3 level is a key parameter for the OH radical atmospheric concentration and for the oxidative and scavenging capacity of the atmosphere to slow down the build-up of greenhouse gases and in particular of CH_4, the most important photochemically dependent greenhouse gas.

We are still far from a complete understanding of this issue so that modeling the climate chemistry response to the emission of anthropogenic pollutants is presently a major challenge.

5.2 ATMOSPHERIC AEROSOLS-CLOUDS

Particulate pollution of the troposphere has long been considered a potentially important climate forcing term (Charlson and Pilat, 1969; Bryson, 1974). Aerosols can modulate the radiative fluxes changing the value of the atmospheric albedo, through direct effect on the scattering and absorption of photons in the short-wavelength and long-wavelength part of the spectrum.

SO_2, when oxidized to sulphate leads to the formation of aerosols. This process may then influence the formation of clouds and their pattern. Similarly the dissolution in the atmospheric water droplets of the oxidation products of SO_2 and NOx, changes the refractive index and then the scattering characteristics of the aerosol.

Major aerosols effects, attributed to anthropogenic pollution, are presently limited to continental areas and several observation studies have suggested that the atmospheric turbidity increased with the industrial activity (Charlson, 1988). Even less is known about a relation between atmospheric chemistry and clouds.

An example of a possible role of NOx in the process of cloud formation can be discussed in the case of the oxidation of dimethylsulphide (DMS). On oceans, an important source of aerosols is believed to be the oxidation to sulphuric and to methanesulphonic acid of DMS, released in large quantities from the biological activity of algae. Sulphuric and methanesulphonic acid are in fact characterized by important aerosol forming properties. If the possible sinks of DMS in air are considered, photolysis, chemical reactions with O_3 or with the radicals OH, NO_3, IO, ClO, BrO, O(3P), it can be seen that only the reactions with OH and NO_3, in turn related to the NOx level in the atmosphere, lead to the formation of sulphuric acid and of methanesulphonic acid. The other reactions in fact seem to have as main reaction product the formation of dimethylsulphoxide, water soluble and probably rapidly scavenged.

In conclusion, while the role of clouds and aerosols in the planetary radiation budget is understood to be of primary importance and recent developments have further increased the importance of the cloud feedback (Ramanathan, 1987; Ramanathan et al., 1989), the impact of anthropogenic trace gases, SO_2, NOx and transformation products, critically affecting the various processes by which H_2O is partitioned into its vapour, liquid and solid phases, is still mostly unknown.

5.3 NITROUS OXIDE

Nitrous oxide, N_2O, a subject somewhat outside of this review, has been recently in the focus of a debate in relation to the evaluation of its emission from coal burning in power plants; for this reason it will be considered here.

Nitrous oxide is important for possible effects on the planetary atmosphere for two reasons:

a. it is practically unreactive in the troposphere with a lifetime of 150 years (Hao et al., 1987) and accumulates in the troposphere at a rate presently evaluated as 0.2-0.3% per year. The heat trapping efficiency per molecule, of N_2O is estimated as large as 250 times that of CO_2;
b. it diffuses to the stratosphere where it reacts with O(1D) atoms to generate NO and trigger the well known O_3 destruction cycle:

$NO + O_3 \longrightarrow NO_2 + O_2$ (12)
$NO_2 + O \longrightarrow NO + O_2$ (49)
Net: $O_3 + O \longrightarrow O_2 + O_2$

A few years ago it was thought that the growth of atmospheric N_2O was predominantly caused by the combustion of fossil fuels, and in particular by coal-fired utility boilers; soil, ground water and oceans are the other important sources of this gas.

This view has been changed dramatically by the discovery of an artifact in the analysis of grab stack samples; laboratory studies have in fact shown (Muzio et al., 1989) that an heterogeneous reaction between NOx and SO_2 in the presence of water was occurring to reduce NOx to N_2O.

At the present state of knowledge it appears that the N_2O emissions from uncontrolled stationary and mobile sources can be considered negligible with respect to the other sources in the global budget. At the same time however, speculations and preliminary experimental data seem to indicate that the N_2O flux from fluidized bed combustion of coal and catalyst controlled gasoline engines might give a non negligible contribution (DeSoete, 1990). Moreover an indirect effect of the anthropogenic pollution by SO_2 and NOx has been recently considered. The biogeochemical cycle of nitrogen is clearly perturbed by increased dry and wet deposition of ammonium and nitrate ions, a source of fixed nitrogen which unbalances that due to natural sources. Nitrifying and denitrifying bacteria would use it as a nitrogen source to produce more N_2O. Estimates of this source (Elkins, 1989) give values of 0.13-5 Tg (N) per year; up to about 1/3 of the total N_2O flux to the atmosphere.

6. Conclusion

Fossil fuel combustion is understood to have important effects on the radiative balance at the planetary scale (CO_2), and to control, via emission of the reactive gases SO_2 and NOx, phenomena such as photooxidant formation, visibility degradation, corrosion of materials and acidity deposition at the local/regional scale in the continental boundary layer of industrialized countries. The impact at the global scale of the injection in the atmosphere of these reactive gases is however less known.

The contribution of the combustion of fossil fuel and of the metallurgical industry to the SO_2 global budget is about one half of the total, with sources very unevenly distributed.

Burning fuels is also the primary anthropogenic source of NOx; its contribution to the global budget is certainly very important, but the balance is less known. Unlike SO_2, NOx emissions are still on a rising trend.

The present knowledge in atmospheric chemistry/photochemistry gives sufficient reliability to most of the processes occuring in the gas phase, the chemistry/photochemistry of the atmospheric water phase is less comprehended as it is the one taking place on solid atmospheric aerosols.

SO_2 and NOx are converted into acids by oxidants through a set of gas phase and aqueous phase reactions, of comparable importance, occurring both in daytime and at night; the formation of the oxidants involved in these processes is largely controlled by the atmospheric chemistry of the oxides of nitrogen and of the anthropogenic and natural volatile organic compounds.

Experimental evidence and model calculations strongly support the view that a reduction in the emission of reactive trace gases by suitable abatement techniques should significantly reduce the adverse phenomena previously indicated.

6.1 SOURCE-RECEPTOR RELATIONSHIP

It has been concluded that a mitigation or a prevention of the effects of pollutants to the environment, requires a decrease in exposure through emission reduction. There has been considerable debate about how much and where reductions of effects can be achieved by SO_2 and NOx emission planning. This problem is generally discussed in terms of "source-receptor relationship", defined as the quantification of the change, including time and space scales, in deposition (wet and dry) at a receptor caused by a change in the emission. The impact of the atmospheric chemistry in the search of a solution to this problem is quite evident.

From the previous discussion, it comes out that the fraction of SO_2 and NOx oxidized to sulphate and nitrate by homogeneous gas phase reaction, in the liquid phase or on surfaces, strongly depend on the particular atmospheric chemical and meteorological conditions encountered. A severe problem lies in the non linear relation between emission rates of primary pollutants and deposition of products formed by the oxidation in the atmosphere of these species. SO_2 is oxidized in the gas phase by the hydroxyl radical and the process is linear (see reactions 23,25). In the aqueous phase SO_2 is known to be oxidized very rapidly, but the rate may be limited by the supply of oxidants, which depends upon gas phase reactions involving VOC and NOx, and by the rate of incorporation of SO_2 into the liquid phase. There is evidence that this process tends to be non linear. Model calculations show that the oxidation of NOx to HNO_3 is quite non linear, in its dependence on the [VOC]/[NOx] ratio and on the [NOx]. The conversion of NOx to NO_3 to HNO_3 is not a linear process either: the rate depends on the concentration of O_3, which tends to increase with the NOx level in the rural environment. Therefore the oxidation becomes faster when more nitrogen oxides are injected into the atmosphere.

Up to now sometimes ambiguous results have been obtained from comparisons between emissions and measurements as alternative to calculations.

What above discussed leads to emphasize that an accurate evaluation of the atmospheric chemistry as well as of transport/meteorology must support the planning and application of SO_2 and NOx abatement techniques.

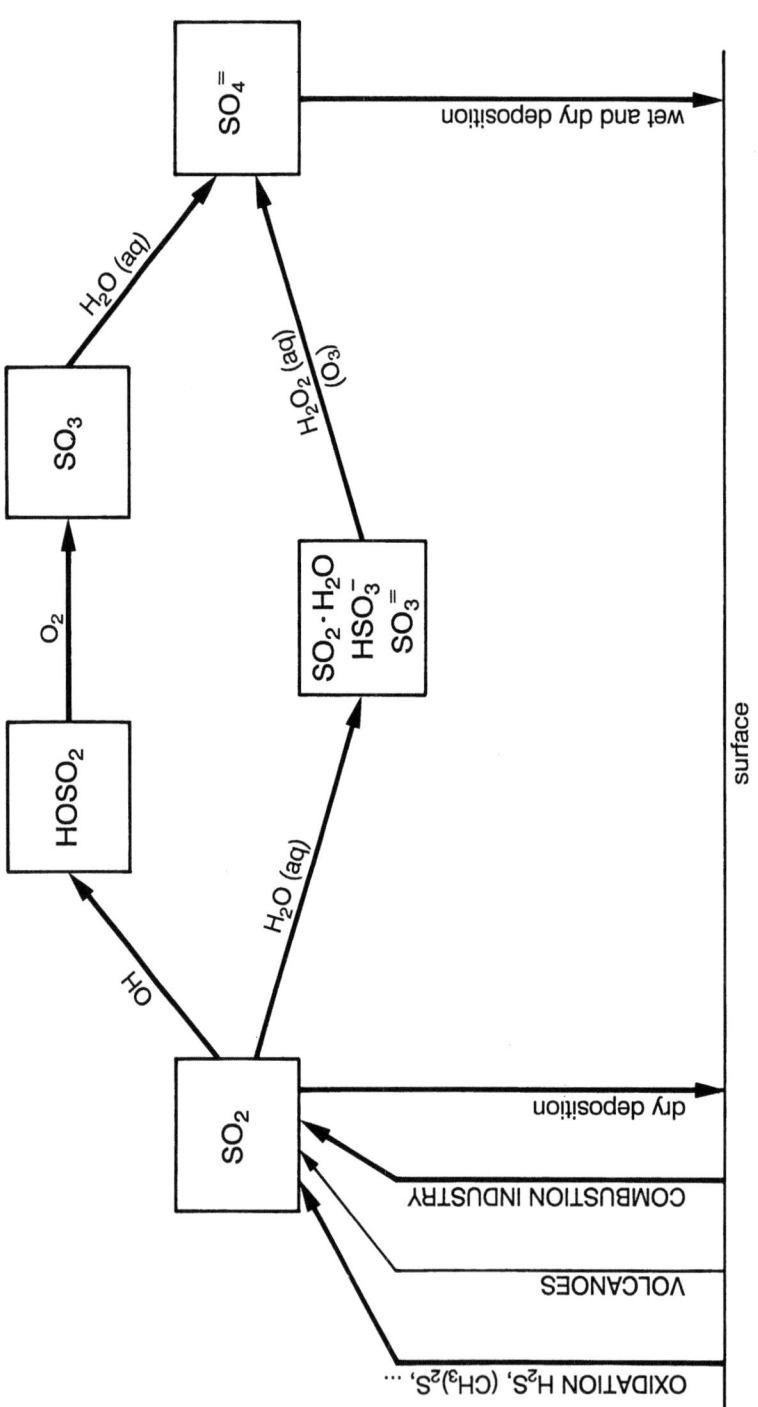

Fig. 1 - Sulphur dioxide: principal homogeneous gas-phase and heterogeneous transformation reactions, products and removal paths.

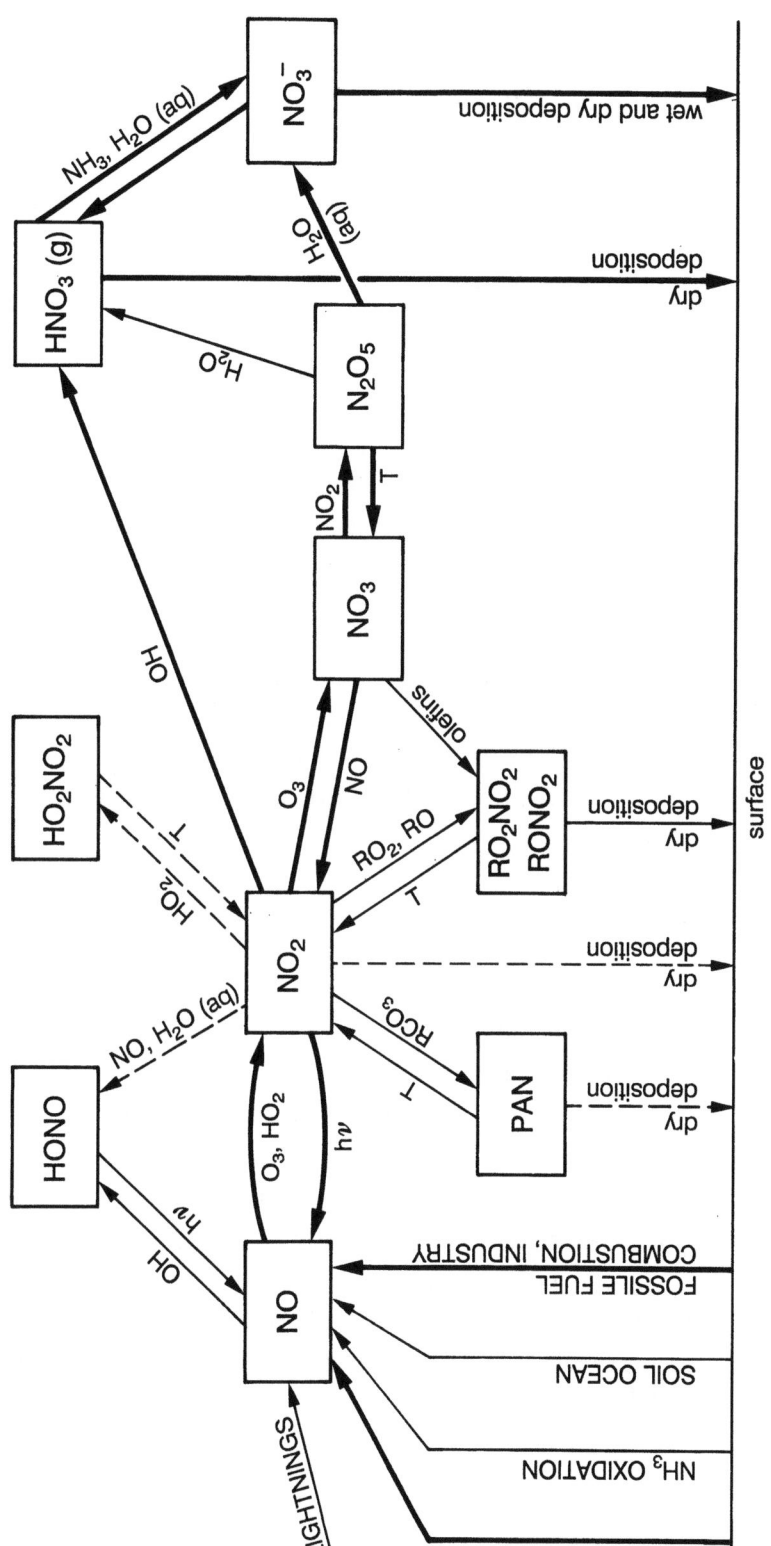

Fig. 2 - Nitrogen oxides: principal homogeneous gas-phase and heterogeneous transformation reactions, products and removal paths.

Acknowledgements

The suggestions and the criticism of J.Hjorth, N.Jensen and H.Skov, are gratefully acknowledged.

References

Atkinson, R. (1990), Atmos. Environ., **24A**, 1.

Atkinson, R. and Lloyd, A.C. (1984), J. Phys. Chem. Ref. Data, **13**, 315.

Barnes, I., Bastian, V., Becker, K.H. and Zhu Tong (1990), **94**, 2413.

Bryson, R.A. (1974), Science, **184**, 753.

Burrows, J.P., Tyndall, G.S., Schneider, W., Bingemer, H., Moortgat, G.K. and Griffith, D.W.T. (1986), NBS Special Publication, **716**, 137.

Calvert, J.G. Ed. (1984), "SO_2, NO and NO_2 Oxidation Mechanisms: Atmospheric Considerations", Butterword, Boston.

Hampson, R.F., Jr. and Garvin, D. (1978), N.B.S. Special Publication Nr. 513.

Calvert, J.G. and Stockwell, W.R. (1984), in "SO_2, NO and NO_2 Oxidation Mechanisms: Atmospheric Considerations", Calvert, J.G. Ed., Butterwords, Boston, p.1.

Cantrell, C.A., Stockwell, W.R., Anderson, L.G., Busarow, K.L., Perner, D., Scheltekopf, A., Calvert, J.G. and Johnston, H.S. (1985), J. Phys. Chem., **89**, 139.

Cantrell, C.A., Davidson, J.A., Busarow, K.L. and Calvert, J.G. (1986), J. Geophys. Res., **91**, 5347.

Chameides, W.L. (1986), J. Geophys. Res., **91** (D50, 5331.

Chang, T.J., Norbeck, J.M. and Weistock, B. (1979), Environ. Sci. Technol., **13**, 1534.

Charlson, R.J. (1988), in "The Changing Atmosphere", Rowland, F.S. and Isaksen, I.S.A. Eds., J.Wiley and Sons, Chichester.

Charlson, R.J. and Pilat, M.J. (1969), J. Appl. Meteor., **8**, 1001.

Cox, R.A. and Penkett, S.A. (1971), Nature, **230**, 321.

Cox, R.A. and Penkett, S.A. (1972), J. Chem. Soc. Faraday Trans., I, **68**, 1735.

Crutzen, P.J. (1979), Annual Review of Earth and Planetary Sciences, 7, 443.

Daubendiek, R.L. and Calvert, J.G. (1975), Environ. Lett., **8**, 103.

Davis, D.D., Ravishankara, A.R. and Fischer, S. (1979), Geophys. Res. Lett., **6** , 113.

DeMore, W.B., Sander, S.P., Golden, Molina, M.J., D.M., Hampson, R.F., Kurylo, M.J., Howard, C.J. and Ravishankara, A.R. (1990), J.P.L. Special Publication 90-91.

DeSoete, G. (1990), paper presented at the "European Workshop on N_2O", LNETI/EPA/IFP, Lisboa, 6-8 June, 1990.

Dickerson, R.R. et al. (1987), Science, **235**, 460.

Elkins, J.W. (1989), Contribution to the Intergovernmental Panel on Climate Change (IPCC), in press.

Finlayson-Pitts, B.J. and Pitts, Jr.,J.N. (1986), Atmospheric Chemistry, J. Wiley and Sons, New York.

Graedel, T.E. and Goldberg, K.I. (1983), J. Geophys. Res., **88**, 10865.

Graham, R.A. and Johnston, H.S. (1978), J. Phys. Chem., **82**, 254.

Grosjean, D. (1983), Envoron. Sci. Technol., **25**, 263.

Herron, J.T., Martinez, R.I. and Huie, R.E. (1982), Int. J. Chem. Kinet., **14**, 201.

Hoffmann, M.R. (1985), Atmos. Environ., **19**, 388.

Hoffmann, M.R. and Calvert, J.G. (1985), "Chemical Transformation Modules for Eulerian Acid Deposition Models. Vol. II, NCAR, Boulder.

Hjorth, J., Ottobrini, G., Cappellani, F. and Restelli, G. (1987), **91**, 1565.

Hjorth, J., Ottobrini, G. and Restelli, G. (1988), J. Phys. Chem., **92**, 2269.

Hjorth, J., Cappellani, F., Nielsen, C.J. and Restelli, G. (1989), J. Phys. Chem., **93**, 5458.

Hjorth, J., Lohse, C., Nielsen, C.J., Skov, H. and Restelli, G. (1990), J. Phys. Chem., in press.

Hoshino, M., Ogata, T., Akimoto, H., Inoue, G., Sakamoki, F. and Okuda, M. (1978), Chem. Phys. Lett., 1367.

Jensen, N., Hjorth, J., Lohse, C., Skov, H. and Restelli, G. (1990), Atmos. Environ. in press

Kotzias, D., Duane, M., Nicollin, B. and Schlitt, H. (1990), in Physico-Chemical Behaviour of Atmospheric Pollutant, Restelli, G. and Angeletti, G. Eds. Kluwer Academic pag. 394

Lee, Y.N. and Schwartz, S.E. (1983) in "Precipitation, Scavenging, Dry Deposition and Resuspension", Pruppacher, H.R.,Semonin, R.G. and Slinn, W.G.N. Eds. Elsevier, New York, p.453.

Leighton, P.A. (1961), Photochemistry of Air Pollution, Academic Press, New York.

Lelieveld, J. and Crutzen, P.J. (1990), Nature, **343**, 227.

Levy, H. (1971), Science, **173**, 141.

Liu, S.C., Trainer, M., Fehsenfeld, F.C., Parrish, D.D., Williams, E.J., Fahey, D.W., Hubler, G. and Murphy, R.C. (1987), J. Geophys. Res. 92, 4191.

Magnotta, F. and Johnston, H.S. (1980), Geophys. Res. Lett., 7, 769.

Margitan, J.J. (1984), J. Phys. Chem., **88**, 3314.

Martin, L.R. (1984), in "SO_2, NO and NO_2 Oxidation Mechanisms: Atmospheric Considerations", Calvert, J.G. Ed., Butterwords, Boston, p.63.

Martin, L.R., Damschen, D.E. and Judeikis, H.S. (1981), Atmos. Environ., **15**, 191.

Martinez, R.I.and Herron, J.T. (1981), J. Environ. Sci., A16, 623.

Martinez, R.I., Herron, J.T. and Huie, R.E. (1981), J. Am. Chem. Soc., **103**, 3807.

Morris, JR., E.D. and Niki, H. (1973), **77**, 1929.

Morris, Jr., E.D. and Niki, H. (1974), J. Phys. Chem., **78**, 1337.

Muzio, L.J. and Kramlich, J.C. (1988), Geophys. Res. Lett., **15**, 1369.

Noxon, J.F., Norton, R.B. and Henderson, W.R. (1978), Geophys. Res. Lett. **5**, 675.

Platt, U., Perner, D., Winer, A.M., Harris, G.W. and Pitts, Jr., J.N. (1980), Geophys. Res. Lett., **7**, 89.

Platt, U., Perner, D., Winer, A.M., Harris, G.W. and Pitts, Jr., J.N. (1980), Geophys. Res. Lett., **7**, 89.

Platt, U., Winer, A.M., Biermann, H.W., Atkinson, R. and Pitts, Jr., J.N. (1984), Environ. Sci. Technol. **18**, 365.

Ramanathan, V. (1987), J. Geophys. Res., **92**, 4075.

Ramanathan, V., Cess, R.D., Harrison, E.F., Minnis, P., Barkstrom, R.B., Ahmad, E. and Hartmann, D. (1989), Science, **243**, 57.

Roberts, J.M. (1990) Atmos. Environ. **24A**, 243.

Schwartz, S.E. (1984), in "SO_2, NO and NO_2 Oxidation Mechanisms: Atmospheric Considerations", Calvert, J.G. Ed., Butterwords, Boston, p. 173.

Seinfeld, J.H. (1985), Atmospheric Chemistry and Physics of Air Pollution, J. Wiley and Sons, New York.

Shepson, P.B., Edney, E.O., Kleindienst, T.E., Pittman, J.H., Namie, G.R. and Cupitt, L.T. (1985), Environ. Sci. Technol., **19**, 1985.

Sidebottom, H.W., Babcock, C.C., Jackson, G.E., Calvert, J.G., Reinhardt, G.W. and Damon, E.K. (1972), Environ. Sci. Technol., **6**, 72.

Singh, H.B. and Hanst, P.L. (1981), Geophys. Res. Lett. **8**, 941.

Spicer, C.W. (1982), Sci. Total Environ., **24**, 183.

Stangl, H., Kotzias, D. and Geiss, F. (1988), Naturwissenschaften, **75**, 42.

Stockwell, W.R. and Calvert, J.G. (1983), Atmos. Environ., **17**, 2231.

Temple, P.J. and Taylor, O.C. (1983), Atmos. Environ., **17**, 1583.

Tuazon, E.C., Atkinson, R., Plum, C.N., Winer, A.M. and Pitts, Jr., J.N. (1983), **10**, 953.

Wayne, R.P., Barnes, I., Biggs, P., Burrows, J.P., Canosa-Mas, C.E., Hjorth, J., Le Bras, G., Moortgat, G., Perner, D., Poulet, G., Restelli, G. and Sidebottom, H. (1991), The Nitrate Radical: Physics, Chemistry and the Atmosphere", Atmos. Environ. in press.

Wayne, R.P. (1985), Chemistry of Atmospheres, Clarendon Press, Oxford.

Wuebbles, D.J., Grant, K.E., Connel, P.S. and Penner, J.E. (1989) J.A.P.C.A., **39**, 22

DISPERSION AND TRANSPORT OF ATMOSPHERIC POLLUTANTS

S. CIESLIK
Environment Institute
Joint Research Centre
21020 Ispra (Varese) - Italy

1. Introduction

Air pollution is often viewed as a recent problem associated with the 20th century. However, smokey chimneys, smoke and toxic vapours have in fact been subject of complaints since the Roman times. Early documents from the first and second century in ancient Palestine show clearly that people reacted strongly against sources of air pollution; the Mishnah laws called for a minimal distance from a pollution source to residential areas.

The systematic use of tall stacks as an alternative method to control air pollution started in the early 60's. At the same time this development was accompanied by:
1) an increase in the rated power output in power plants from tens to hundreds or thousands of MW;
2) a consolidation of the emission of several processes into one stack;
3) the use of lower quality fuels.

In either case this translated into a significant increase in the emissions from these sources. Tall stacks increase the distance at which the fall-out may take place, from local to regional and continental scale. Today also in remote areas, some pollutants and particularly photooxidants are easily measured.

It appears clearly from such considerations that the transport of pollutants through the atmosphere is an important aspect of environmental protection studies.

In this chapter we will present the physical mechanisms that govern the dispersion and transport of air pollutant; the influence of the state of the "carrying fluid", or, in other words, the role of meteorology; finally, the different techniques of assessing the process will be outlined.

2. Physical mechanisms

2.1. FATE OF AN AIR POLLUTANT

This section is a general overview of what happens when a pollutant is released into the atmosphere.

For methodological purposes, environmentalists usually divide the environment in so-called "reservoirs", which behave like boxes, into which a given pollutant enters (input or source), through which it migrates (transport), and from which it is finally removed (output or sink). This subdivision is somewhat arbitrary, but it helps understanding betters the processes.

Biosphere (vegetation and humus layer), hydrosphere (oceans), sediments and atmosphere are the most commonly used reservoirs. This chapter is obviously devoted to the atmospheric reservoir.

The inputs of pollutants into the atmosphere are mainly:
- point sources (industrial stacks), characterised by an emission flux expressed in mass per unit time (e.g. kg s^{-1});
- line sources (cars along a road), where the fluxes are in kg s^{-1} m^{-1};
- area sources (e.g. heating of houses in a city), expressed in kg s^{-1} m^{-2}.

Once the pollutant has been released, it undergoes various physical and chemical processes: dispersion, transport, chemical transformation, dissolution in cloud droplets, etc. The physical processes, i.e. dispersion and turbulence, are related to atmospheric dynamics, and their mechanisms will be explained with more detail in the next two sections. The chemical transformations are discussed elsewhere in this book; we may here briefly mention that they can be of homogeneous (e.g. purely gas-phase) or heterogeneous (e.g. gas-phase to cloud droplet or gas-phase to aerosol particle). This topic is related to the photochemical smog issue, as well as to the acid precipitation problem.

The removal (or sink) of the pollutants from the atmosphere takes place through the deposition processes to the surface. These are essentially of two types:
- dry deposition, by which gas-phase or particulate pollutants are directly brought in contact with the surface by turbulence or gravitational settling and absorbed on it;
- wet deposition, when pollutants are dissolved in fog droplets or raindrops falling on the surface.

For more details on deposition processes, see e.g. Gaffney et al. (1987), and Hosker and Lindberg (1982).

The typical behaviour of an air pollutant in the atmosphere is shown schematically in Fig. 1.

2.2. TURBULENCE AND DISPERSION

When a given gas is released in the atmosphere, e.g. from a stack, we can see that it generally forms a cloud (called plume), which is expanding with time, while being driven away by the wind. The expansion means that a given quantity of pollutant is contained within a growing volume of air, resulting in a drop of the concentration. That means that the gas dilutes with time, undergoing "dispersion".

One could imagine a single laboratory experiment, in which a gas is kept immobile within a closed volume, e.g. by avoiding local temperature gradients. If we introduce a foreign gas in the tank, the two gases progressively mix up together until any concentration gradient has disappeared.

That phenomenon occurs even in the absence of any apparent motion of the gas and is called molecular diffusion. In such a case, the flux of the diffusing gas, expressed in kg m^{-2} s^{-1}, is expressed by Fick's diffusion law,

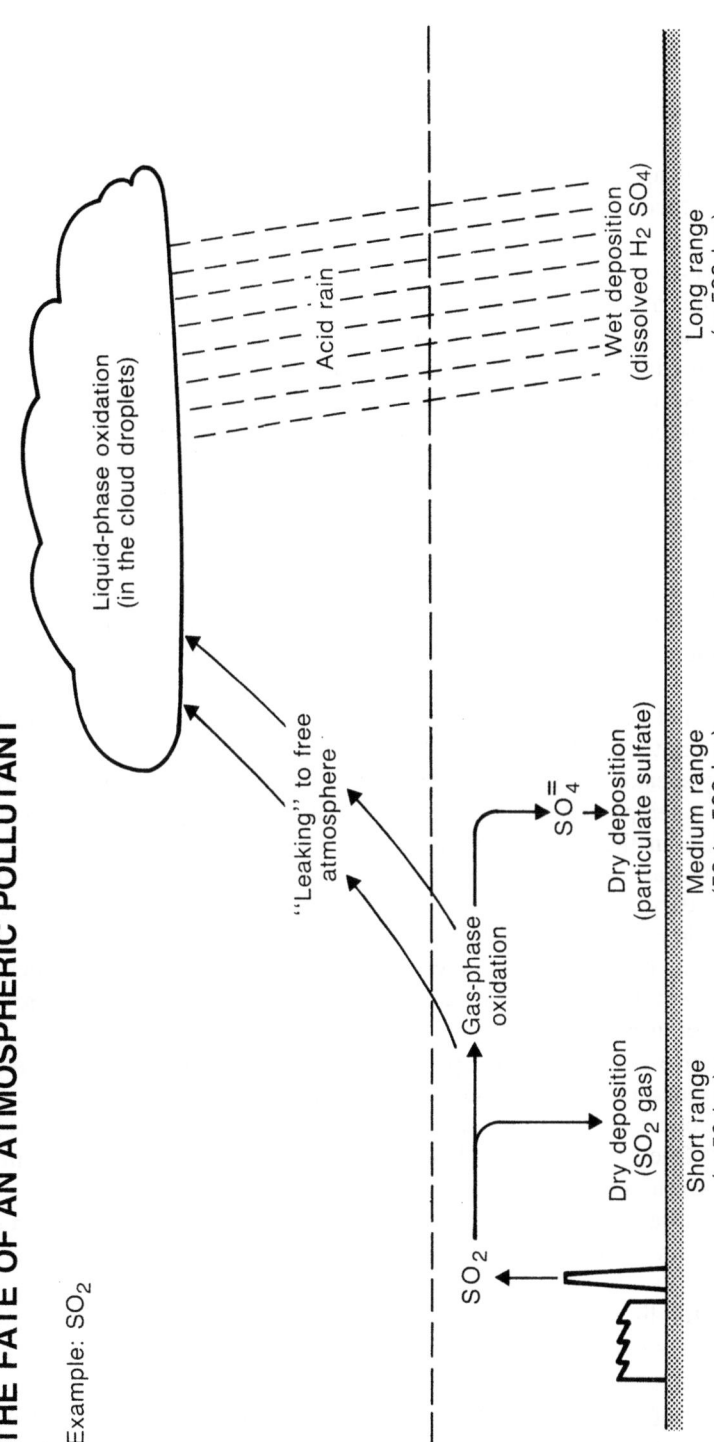

Fig. 1 - Schematic representation of the fate of SO_2, considered as a typical air pollutant, after its release into the atmosphere.

$$\vec{\Phi} = -k \, \overrightarrow{\text{grad} \, C}, \qquad (1)$$

where C is the concentration and k is the diffusion coefficient, which depends on the nature of the gases involved.

However, if we apply the Fickian law to the case of the atmosphere, we see that it is far from being sufficient for explaining the rate at which pollutants are mixed up with surrounding air.

The description becomes more satisfactory if we take into account another phenomenon which is of fundamental importance in atmospheric physics, i.e. turbulence.

Turbulence can be defined as the bulk of the random motions of a fluid. It generally appears as eddies of various sizes and is mainly generated by:
- a mechanical cause: friction of the fluid when it passes over a rough surface;
- a thermodynamical cause, when thermal gradients within the fluid generate convection.

These two conditions are met in the lower layers of the atmosphere: the surface is covered by a variety of roughness elements: vegetation, hills, buildings; the heating of the surface due to absorption of solar radiation generates convective motion, resulting in turbulence.

The part of the atmosphere where these phenomena due to the presence of the surface take place, is called the atmospheric boundary layer (ABL), where turbulence is generally well developed. Above the ABL we find the free atmosphere, where the air flow is usually laminar.

A more detailed description of the meteorology of the ABL is given in section 3. Let us now come back to the phenomenon of turbulence.

Qualitatively speaking, a fluid is considered turbulent when the physical variables characterising it (i.e. temperature, density, etc.) have an apparently random behaviour, as illustrated on Fig. 2. In the absence of turbulence, when the fluid flow is laminar, the values of all the physical variables can be predicted by the equations of fluid mechanics, if the initial and boundary conditions are known. This should theoretically also be true for a turbulent fluid, but in reality the models based on the equations of fluid mechanics fail to predict the behaviour of the fluid. However, they generally succeed if we replace the instantaneous values of the variables by their time averages. There is thus an unexplained aspect in the study of turbulence.

In other to give a more intuitive picture of the phenomenon, we will describe a laboratory experiment aimed at visualising the onset of turbulence. This experiment (see e.g. Swinney and Gollub, 1978) consists of a water tank wherein a foreign coloured liquid is injected (Fig. 3). One sees that the dribble frist moves laminarly, then divides into two dribbles (eddies), which in turn break down into smaller ones.

At a certain distance from the inlet, we can see that the motion liquid has a disorderly appearance, with a number of whirls or eddies of various sizes (developed turbulence).

At even greater distances from the inlet, full mixing occurs between the two liquids, leading to the same result as for molecular diffusion, the difference being that full mixing occurs much more rapidly than in the former case.

Turbulent mixing occurs through the generation of successive eddies, becoming smaller and smaller, each new eddy breaking down into two or more smaller eddies. At the end of the chain, when the size of the eddies becomes of the order of magnitude of

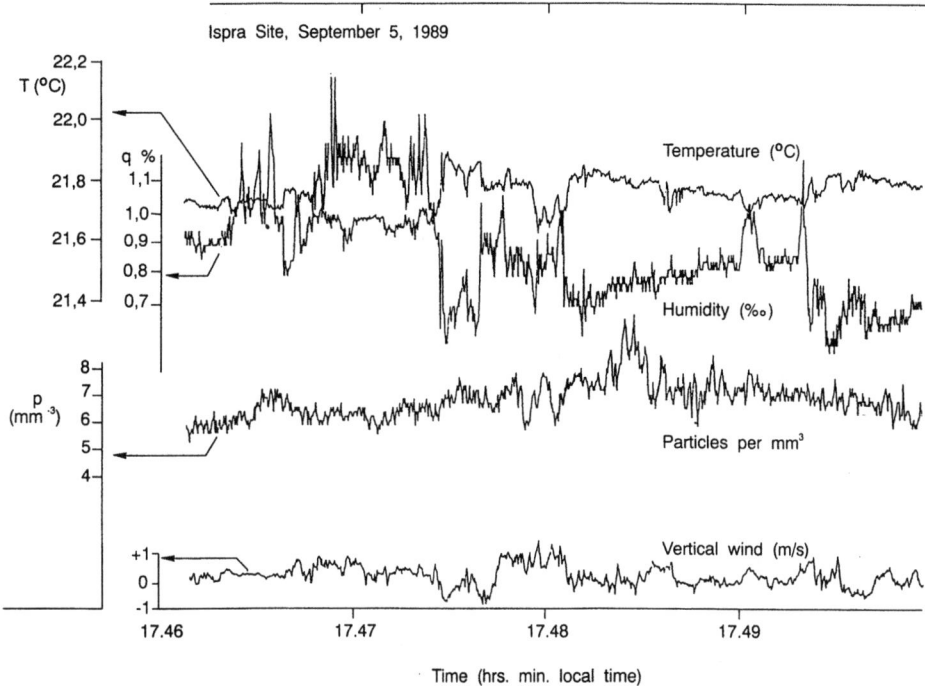

Fig. 2 - An example of random-like behaviour of four atmospheric variables due to turbulence: temperature, humidity, aerosol particles concentration, and vertical component of the wind vector.

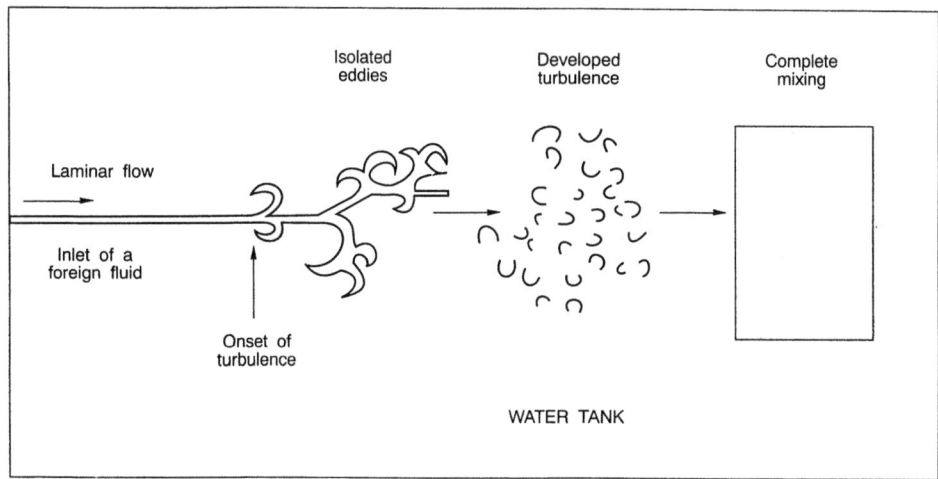

Fig. 3 - Transition to turbulence in a water tank experiment, where a foreign fluid is introduced. The final state is complete mixing.

the molecular mean free path, full mixing occurs.

In energetic terms, kinetic energy is brought into the system by the laminar flow, then distributed among the progressively smaller eddies and finally converted to heat (or enthalphy) by viscous dissipation at the end of the chain.

It has to be noted that the expression "turbulent diffusion", though frequently found in the literature, is not rigorously correct. The only true diffusion process is due to the randomly occurring molecular collisions between molecules of different species. It is more correct to speak of "turbulent mixing". These two processes have a certain apparent analogy, however, that can be shown by the following reasoning.

First we write down the conservation equation for any scalar quantity within a fluid say, the concentration C:

$$\frac{\partial C}{\partial t} + \vec{v} \cdot \overrightarrow{\text{grad } C} = \text{div } \Phi \qquad (2)$$

where v is the velocity vector describing the motion of the fluid, and t is the time. This classical equation simply expresses that the balance of the diffusing substance within a volume element is equal to its flux divergence (=what enters the element minus what comes out) through the frontiers of the element.

If we take Fick's law into account (e.g. 1) and adopt the matrix notation, then eq. (2) becomes

$$\frac{\partial C}{\partial t} + u_i \frac{\partial C}{\partial x_i} = \frac{\partial}{\partial x_i} k \frac{\partial C}{\partial x_i} + S \qquad (3)$$

where the x_i's are equal to the x-, y- and z- coordinates for i=1,2,3, respectively; the u_i are then the three components of the wind vector v; S is the source term. Note that eq.(3) makes use of the Einstein convention, which states that a repeated index within a single term means summation over all the possible values of that index (1,2,3).

We now make use of the Reynolds rule, which is a first attempt to parameterize turbulence by splitting any fluctuating variable in two terms:

$$u_i = \bar{u}_i + u_i'$$
$$C = \bar{C} + C' \qquad (4)$$

where the overbar corresponds to an average over a certain interval which has to be larger than the characteristic time of turbulent fluctuations. Typically, this averaging interval ranges from 10 to 30 min.

Within the boundary layer, air behaves as an incompressible fluid, which is expressed by:

$$\partial C' \frac{\partial u_i'}{\partial x_i} = 0 \qquad (5)$$

taking (4) and (5) into account, the left-hand side of eq.(3) can be rewritten

$$\frac{\partial \overline{C}}{\partial t} + \overline{u}_i \frac{\overline{\partial C'}}{\partial x_i} + \overline{u'_i \frac{\partial C'}{\partial x_i}} = \frac{\partial \overline{C}}{\partial t} + \overline{u}_i \frac{\partial \overline{C}}{\partial x_i} + \frac{\partial \overline{u'_i c'}}{\partial x_i} \qquad (6)$$

and eq.(3) becomes:

$$\frac{\partial \overline{C}}{\partial t} + \overline{u}_i \frac{\partial \overline{C}}{\partial x_i} = \frac{\partial}{\partial x_i}\left[k \frac{\partial \overline{C}}{\partial x_i} - \overline{u'_i c'} \right] + S \qquad (7)$$

This formulation is similar to that of eq.(3), but the dependent variable is no more the instantaneous concentration, but the time averaged concentration. Further, a new term appears, which contains the contribution of the turbulent fluctuations and has the dimensions of a flux. In order to make that term formally similar to the molecular diffusion term, it is convenient to introduce the following parameterisation:

$$\overline{u'_i c'} = -K_c \frac{\partial \overline{C}}{\partial x_i} \qquad (8)$$

where K_c is called "turbulent diffusion coefficient". Substituting this expression of the turbulent flux in (7), we obtain the final form of the conservation equation for a chemical species within a carrier gas:

$$\frac{\partial \overline{C}}{\partial t} + \overline{u}_i \frac{\partial \overline{C}}{\partial x_i} = \frac{\partial}{\partial x_i} (k+K_c) \frac{\partial \overline{C}}{\partial x_i} + S \qquad (9)$$

Now the formal analogy with eq.(3) is perfect if we replace the instantaneous values by their averages. The fluctuations are parameterised by the coefficient K_c. Furthermore this coefficient K_c is generally several orders of magnitude greater than k and the latter can thus be neglected.

As a conclusion of this section, we can state that in a turbulent fluid, a diffusing gas behaves "as if" turbulence caused an additional diffusion, much more efficient than the molecular one, this phenomenon being described by a conservation equation where the dependent variable is the time averaged concentration.

2.3. TRANSPORT

The "transport" of the pollutant plume is a process that is qualitatively distinct from diffusion or dispersion. Roughly speaking, it is the bulk horizontal displacement of the plume, carried by the wind. Mathematically, the bulk transport, also called advection, is expressed by the term $\overline{u}_i(\partial \overline{C}/\partial x_i)$ which appears in the continuity equation.

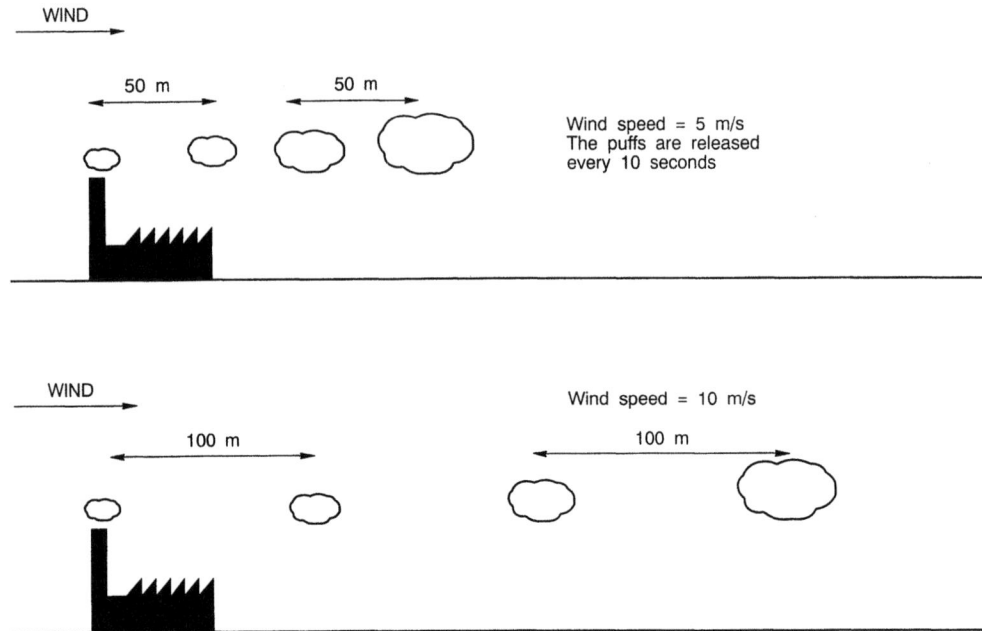

The higher wind speeds, the greater the distance between the puffs

Fig. 4 - If smoke puffs are released at successive times, the difference between dispersion (causing the expansion of the cloud) and transport (which is the horizontal displacement of the cloud) appears clearly. Note that increasing wind speeds result in diluting the pollutants. This effect remains if we replace the separated puffs by a continuous release.

The difference between dispersion and advection is illustrated on Fig. 4; dispersion expands the plume, where concentration decreases with time. Advection takes it away over longer distances, having only an indirect effect on concentration, which is lower if the wind speed is higher.

3. Meteorology and air pollution

Meteorology has an obvious importance in the study of the behaviour of air pollutants. Its influence can be of various kinds: wind, synoptic situation, role of the planetary boundary layer, atmospheric stability. We will successively examine these aspects.

3.1 WIND SPEED AND DIRECTION

The effect of wind speed on air pollutant concentration is illustrated on Fig. 4; in a plume, the concentration is roughly inversely proportional to the wind speed.
It is well known that severe pollution episodes occur during calm wind periods.
More generally, wind speed and direction obviously govern the trajectories of pollutants during short-, medium- and long range transport.

The knowledge of the three-dimensional wind vector field is thus of crucial importance in any attempt of assessing pollutant trajectories. The wind field is a consequence of the synoptic situation, and, more specifically, of the horizontal pressure gradient. This is valid only for the geostrophic wind, which is a rather theoretical concept resulting from the assumed dynamic equilibrium between the forces applied on an air parcel within the pressure field (see Fig. 5). The geostrophic wind represents a close approximation to the real wind in the free atmosphere, whereas the approximation is no more valid for the boundary layer. Its direction is parallel to the isobars, and follows the Buys-Ballot law, which states that, if we look towards the direction of the wind vector, the higher pressures are situated on the right side in the Northern Hemisphere. In the Southern hemisphere, the opposite law holds.

The equilibrium state which causes the geostrophic wind is the resultant of the pressure gradient induced force and the Coriolis force; the magnitude of the gastrophic wind vector is given by:

$$V_g = \frac{\partial p/\partial x}{2 \rho \Omega \sin \varphi} \tag{10}$$

where $\partial p/\partial x$ is the horizontal pressure gradient, ρ is the air density, Ω is the angular velocity of earth rotation, and φ is the latitude.

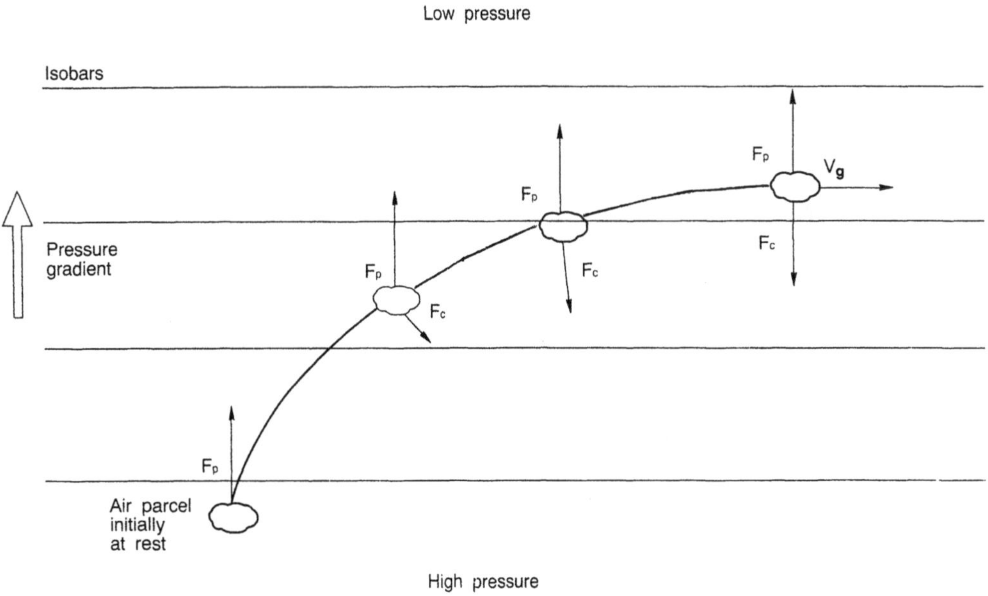

Fig. 5 - Air parcel responding to the pressure-gradient force F_p. If the parcel is initially at rest, F_p causes it to accelerate, but the Coriolis force F_c deflects it until it becomes equal but opposite to F_p. That equilibrium state is the geostrophic flow, represented by the vector V_g.

In the boundary layer, the situation is rather different. The surface roughness causes a friction effect which introduces an additional force. This modifies the wind field as one approaches the surface. The direction is modified, giving rise to the "Ekman spiral" (see Fig. 6); the wind speed decreases as the altitude decreases and theoretically vanishes at the surface level. In the ideal case of a neutral atmosphere, the wind speed u follows the logarithmic law:

$$u = \frac{u_*}{k} \ln\left(\frac{z}{z_0}\right) \tag{11}$$

where u_* is the friction velocity, k is the von Karman constant, z is the height a.g.l. (above ground level) and z_0 is the roughness length, depending on the size of the surface roughness elements: hills, trees, buildings. Generally, z_0 is about one tenth of the mean height of the roughness elements. Table 1 gives some typical values.

The friction velocity is a concept of the theory of turbulence. It is given by:

$$u_* = \sqrt{-\overline{u'w'}} \tag{12}$$

where u' and w' are the fluctuations of the horizontal and vertical components of the wind vector, respectively.

There are other formulations of the vertical profile of the wind speed for various situations, and they can be found in any textbook on the subject (see e.g. Stull (1988)), or in more specific publications (Paulson (1970)).

The important things to remember is that the wind speed varies considerably throughout the boundary layer and that it is higher at upper levels.

The shape of the wind speed profile depends on another factor, which will be dealt with in the following section: the atmospheric stability.

Table 1 - Some values of the roughness lenght z_0 for different types of surface.

Type of surface	z_0 (cm)
Ice (flat)	0.001
Snow	0.005 - 0.01
Smooth desert	0.03
Short grass	0.3 - 1.0
Long grass	4 - 10
Agricultural crops	4 - 20
Forests	100 - 600
Buildings, cities	200 - 5000

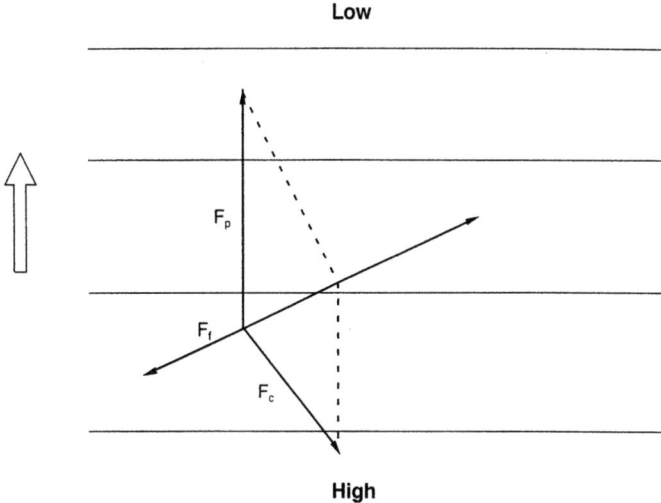

Fig. 6a - The friction due to the presence of the surface causes and additional force F_f to deviate the air flow from the geostrophic wind direction.

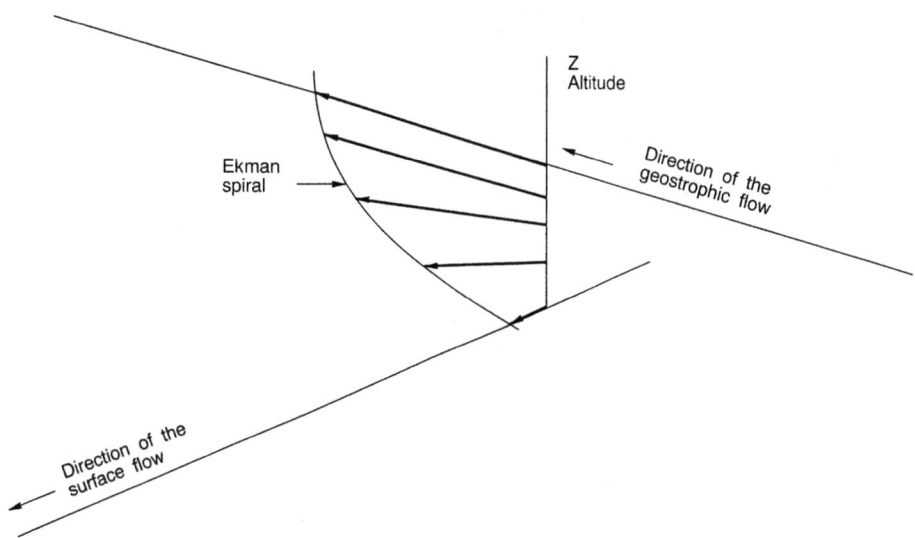

Fig. 6b - The wind direction generally changes with altitude in the boundary layer (Ekman spiral). In the free atmosphere, it is equal to the geostrophic wind direction; at lower altitudes, the frictional force progressively changes the direction.

3.2 ATMOSPHERIC STABILITY

The concept of stability is of crucial importance in air pollution studies since it controls the dispersive capacity of the atmosphere.

It must be understood under its mechanical meaning. In any mechanical system in equilibrium, the stability is the ability of that system to return to its equilibrium state after having been deviated from it by any cause. In the case of the lower atmosphere, the equilibrium which is referred to is also based on thermodynamical considerations: if we leave the atmosphere, i.e. a perfect gas in a gravitational field, with all the forces being in mechanical equilibrium, and if we solve the thermodynamical equations, we obtain the adiabatic vertical temperature profile:

$$T = T_o \left(\frac{p}{p_o}\right)^{R/C_p} \qquad (13)$$

where T and p are the temperature and the pressure; the subscript o refers to a certain reference level (e.g. where p = 1000 mb); R is the constant of the perfect gases and C_p is the air heat capacity. The vertical temperature gradient related to that profile is called "adiabatic lapse rate" and is often represented by the symbol Γ.

In the lowest atmospheric layers, we can adopt:

$$\Gamma = 9.8 \times 10^{-3} \text{ K.m}^{-1} \qquad (14)$$

Three cases are now possible:
1) $\partial T/\partial z > \Gamma$
   In this case, if any cause brings an air parcel upwards adiabatically, its temperature changes following the law (13); it will then be cooler than surrounding air, thus denser; consequently, it will "fall" and go back to its original level. This is thus the *stable case*.
2) $\partial T/\partial z < \Gamma$
   This is the opposite situation, and a reasoning similar to the previous one leads to the conclusion that the air parcel brought upwards is brought even farther upwards. This is the unstable case.
3) $\partial T/\partial z = \Gamma$
   Here, an air parcel brought upwards will remain at its new level. We are now in presence of the *neutral case*.

As a consequence of these properties, the stability of the atmosphere may also be described as its ability to suppress or enhance vertical motion.

Vertical motion can primarily be caused by mechanical factors, like slopes that force the air to go up or down, and roughness elements that cause mechanically driven eddies. It can also be caused by any source of heat that results in buoyancy forces. An unstable atmosphere enhances vertical motion; a stable atmosphere suppresses it.

The effect of stability on the dispersion of air pollutants appears as a consequence of its effect on vertical motion. Taking the example of a single stack plume, instability cause important vertical spreading, which dilutes the pollutants, but bring them rapidly in contact with the ground level. On the contrary, a stable atmosphere causes narrow, higly concentrated plumes, which can "travel" over long distances without being in contact with the surface.

Because of its importance, the stability of the boundary layer has been studied extensively with the aim of developing methods of evaluating it.

The most direct method is the measurement of the vertical temperature profile; this requires the presence of tall meteorological towers (more than 100 m high) or remotely sensed observation systems. These are always costly and thus rather uncommon.

Air pollution specialists developed thus readily usable parameterisations based on surface observations.

On of the mostly used systems is the Pasquill classification scheme, where the stability is described by a set of categories defined by very simple and readily observable data: the cloud cover and the wind speed (Pasquill (1961)).

The importance of the sky cover appears if one reminds that the amount of solar radiation reaching the surface acts on the vertical temperature profile.

The role of wind speed appears less clearly. It is related to the ability of wind to generate turbulence. The stronger the wind, the intenser the mechanical turbulence due to the friction of the wind at the surface. The effect of that kind of turbulence is the mixing up throughout the depth of the boundary layer. A vertically fully mixed atmosphere is neutral; hence strong winds favour neutrality, destroys vertical gradients, thus suppressing both stable and unstable states.

The Pasquill classification, consisting of categories referred to by the letters A thought F, is shown on Table 2.

Table 2 - The Pasquill stability classification scheme:
A means very unstable; B means unstable; C means slightly unstable; D means neutral; F means stable; G means very stable.

Insolation/Cloud Cover		Surface Wind Speed (m/s)				
		<2.0	2 to <3	3 to <5	5 to <6	$\geq$6
Day	Strong Insolation	A	A-B	B	C	C
	Moderate Insolation	A-B	B	B-C	C-D	D
	Slight Insolation	B	C	C	D	D
Day or Night	Overcast	D	D	D	D	D
Night	Thin overcast or $\geq$0.5 cloud cover	-	E	D	D	D
	$\leq$0.4 cloud cover	-	F	E	D	D

There are other stability schemes or parameters, but their full description is beyond the scope of this presentation. See e.g. Pasquill, 1974.

Nevertheless it is worthwhile to mention an important quantitative expression of the atmospheric stability, the Richardson number, expressed by:

$$Ri = \frac{g}{T} \frac{\partial \theta / \partial z}{(\partial u / \partial z)^2} \qquad (15)$$

where both the vertical gradients of temperature and wind speed appear explicitly.

3.3 MIXING LAYER AND INVERSIONS

The vertical temperature gradient is not uniform with height, nor constant with time. The boundary layer may have a layered structure, with stable and unstable layers superimposed.

During a sunny day, the boundary layer is generally unstable, but at a certain altitude, the temperature gradient changes sign and becomes positive (elevated inversion layer). The consequence of this kind of vertical structure on pollutant dispersal (Fig. 7) is important, since instability favours vertical spreading of the pollutants, while stability inhibits it. Hence, during a day when such a structure is observed, air pollutants are mixed up throughout the unstable layer (also called mixed or mixing layer), but their vertical

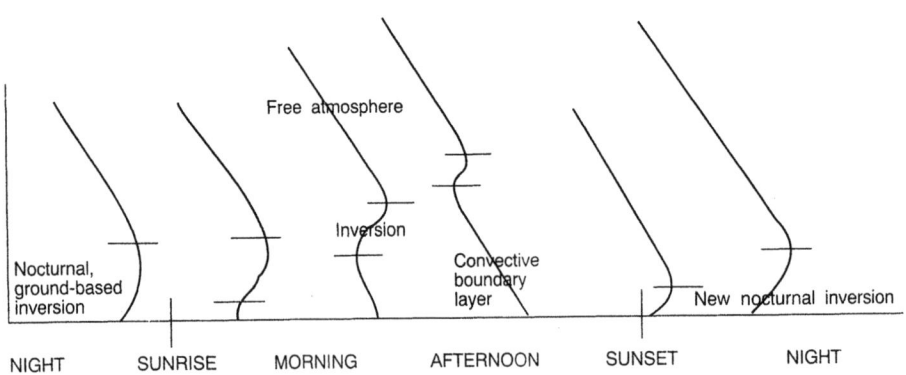

Fig. 7 - Typical evolution of the vertical temperature profile during a cloudless 24-h period.

diffusion is limited by the elevated inversion layer. Further within the mixed layer, the concentration of the pollutants is lower if the height of that layer is greater. This height (called mixing height or mixing depth), which is defined as the height where the first inversion layer begins, is thus an important pollution controlling parameter, that can be measured by balloon soundings or by remote sensing acoustic systems (acoustic radars); Fig. 8 illustrates an acoustic radar recording.

The vertical structure described above is far from being the only picture of the atmospheric boundary layer. This structure varies with time and depends closely on the radiative budget of the earth surface.

Heating and cooling of the surface depend on whether soil is absorbing more radiative energy than it looses, or the opposite. Absorption of radiation is possible when the sun shines (clear sky); the soil emits continuously infrared radiation, due to its blackbody emission. During sunny periods, absorption of solar radiation exceeds emission, and the soil is heated.

During a cloudless night, the soil is cooled by loss of radiative energy. If the sky is covered by clouds, a radiative equilibrium is established between upward radiation from the soil and downward radiation from the clouds. The radiation budget at ground level is then close to zero, and practically nor heating nor cooling occurs at the surface.

The importance of the cooling and heating processes at the surface appears clearly if one keeps in mind that the atmosphere is practically transparent to electromagnetic radiation in a large spectral range. Radiation is thus largely unable (with the exception of the infrared) to result in temperature changes of air; thus, the temperature variations occurring at ground level are, roughly speaking, the only cause of heating and cooling in the lower atmosphere. This process occurs by propagation of the temperature changes by turbulence towards higher altitudes (remind that conductivity is insufficient for propagating heat in air).

Taking these considerations into account, the evolution of the boundary layer during a typical cloudless 24-hour period can be described as follows:

After sunset, cooling of the soil starts up, as stated before, and is cooling propagates upwards through the air. The result is the progressive formation of a "ground-based inversion layer", i.e. a layer where the vertical temperature gradient is strongly positive. The boundary layer is thus generally stable during cloudless nights; at mid-latitudes, a nighttime inversion is typically a few hundreds of meters deep.

After sunrise, the soil heats up rapidly, and the inversion is progressively destroyed and "pushed" upwards by the new unstable mixed layer. This evolution is shown in Fig. 7. If cloud cover is present, the existence of a quasi-equilibrium of the radiation field results in a nearly neutral atmosphere throughout the day-night cycle.

These patterns are modified in the case of disturbed topography, or in the presence of shorelines that generate breeze circulations.

3.4 THE ROLE OF ATMOSPHERIC WATER

Water is always present in the atmosphere under whatever physical state: vapour (humidity), liquid (fog or cloud droplets, raindrops), solid (ice crystal, snow).

Water vapour has some indirect effects on air pollution by modifying the stability of the atmosphere, owing to its property of "infrared radiator". We indicated previously

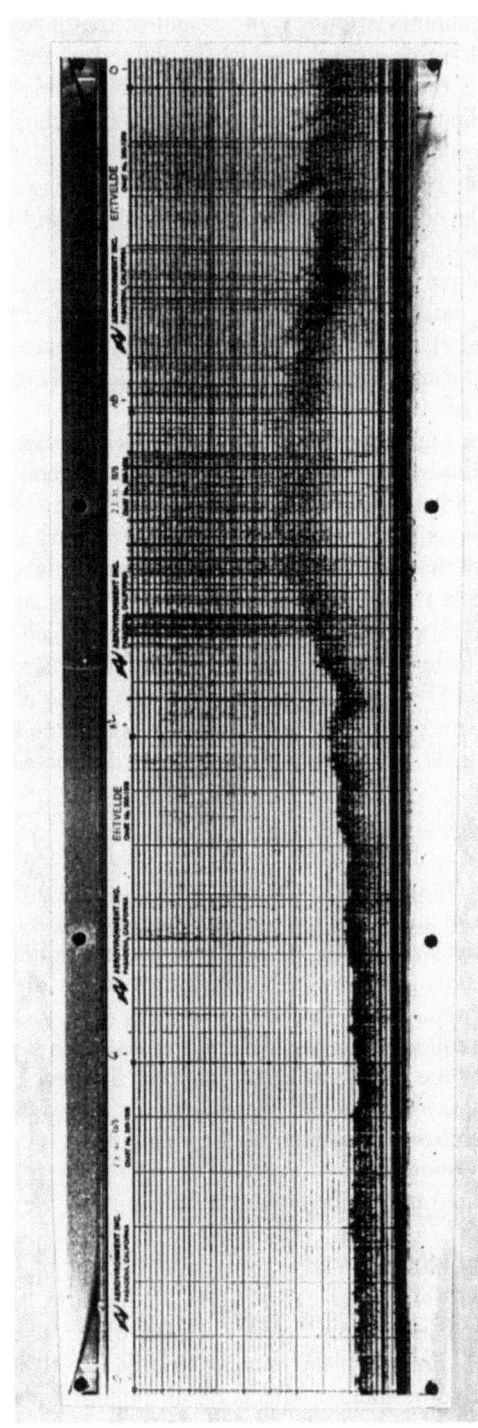

Fig. 8 - An example of the 24-h record of the boundary layer structure observed by an acoustic radar. The black zones corresponds to the inversion layer.

that the atmosphere is transparent to radiation. In the case of a wet atmosphere, this is true only in first approximation. It is known from spectroscopy that the water molecule absorbs and emits infrared radiation; since a part of that radiation is situated in a spectral range corresponding to the blackbody radiation around 300 K; that means that the presence of a high concentration of water vapour results in heat exchange and thus in modification of the vertical temperature gradients.

The presence of water vapour has also an effect on the size of particulate matter, including pollutant aerosols. Liquid and solid atmospheric water plays an important role in air pollution transport. Soluble pollutants can be dissolved in fog and cloud droplets. Fog may be the carrier of strong acid deposition episodes, since its droplets are often very little-sized and formed around a solid pollutant particle (e.g. metal salts, or even droplets of sulfuric oleum due to the oxidation of SO_2); the concentration of pollutants, especially acid, can then reach very high values. Deposition occurs when turbulence brings the droplets in contact with the surface. This phenomenon is particularly marked in elevated areas covered with forest, where fog is frequent. A typical case is the Black Forest in Southern Germany.

Pollution of cloud droplets represents another type of problem. Clouds may travel over very long distances before being converted into rain. This is thus associated with long-range transport. Further, the long times associated with that process are sufficient for chemical conversions to take place. If oxidants (ozone, hydrogen peroxides, ferrous salts) are present in the aqueous phase, then sulphur and nitrogen dioxide are converted into sulfuric and nitric acids. So, when the cloud precipitates, the resulting rain is acid. This is, in few words, the origin of the acid rain problem.

4. Assessment techniques

The assessment of air pollution dispersion and transport is not a simple task. No method permits to follow accurately and unequivocally a given puff of pollutant molecules and to determine exactly what happens with them as long as they travel through the atmosphere. Most methods are more or less indirect: concentrations measurements, modelling meteorological observations, tracer releases.

4.1 CONCENTRATIONS MEASUREMENTS

The classical way for controlling the air quality in a polluted area, industrial or urban, is a network of stations equipped with meteorological and chemical sensors. Data available every hour (or half-an-hour) are transmitted to a control station, for a synoptic view, on which decision-makers can ask in extreme cases for a reduction of emissions. As mentioned above, meteorological parameters are related to pollutant concentration. This is the reason why every pollution monitoring station must be complemented by a meteorological station, where several parameters must be recorded (wind speed and direction, radiation).

The positioning and density of monitoring stations should provide a true representation of the receptor area, taking into account emission source characteristics, local orography and meteorology, and the distribution of population. Actually siting and density of monitoring stations are quite different, from the equally spaced grid pattern of the

Dutch national network to the other extreme of few stations in downtown. The measured parameters are different according to local emission; also reference methods may differ from one country to another. If time series of pollution and meteorological data are available, it is possible to predict the pollution level for the next day by a statistical model and meteorological forecast.

In fact, the information given by these monitoring networks on pollution dispersion and transport is only indirect, and their are mainly used as a supporting tool for decision-makers (design of regional abatement regulations; special measures to be taken in cases of pollution episodes).

A better picture can be obtained with the use of mobile measuring units, which are able to follow pollutant plumes (see e.g. Sandroni and Cerutti, (1989)) or to record concentration profiles along given horizontal lines or contours (Fig. 9).

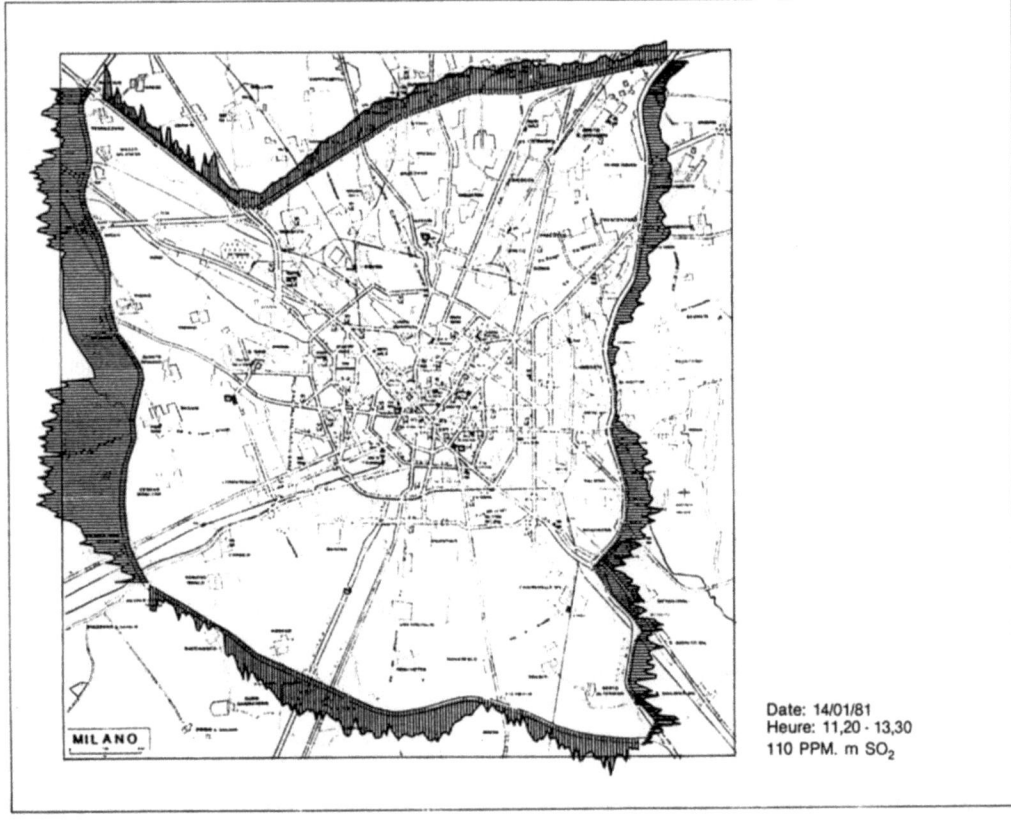

Fig. 9 - Mobile measuring units are a powerful means of assessing air pollution along a given line. Here on instrumented van measured continuously the SO_2 burden around the city of Milan.

4.2 MODELLING

The remarkable development of computer technology during the last three decades made the numerical simulation of physical phenomena easier and more powerful.

In the case of dispersion and transport of air pollutants, a huge amount of numerical models exist, ranging from very simple ones to very complex codes that reach a high degree of sophistication. Some of them are commercially available.

Atmospheric modelling is a practically self-contained topic that generated an abundant literature (see e.g. Pielke, 1984 and references therein).

This section will thus represent an overview of different types of models.

Models can be:
- deterministic or statistical;
- numerical or analytical;
- eulerian or lagrangian.

They can be time-dependent or stationary, diagnostic or prognostic; they can take chemical processes into account or not; they may be 1-, 2- or 3-dimensional.

A *deterministic model* is obviously based on the general deterministic principle of physics, which states that the state of a system at a time t_1 is fully determined by the state of that system at a previous time t_2. The two states are related by an equation of evolution, which is generally written in differential form and expresses the physical law that governs the behaviour of the system.

In our case, the evolution equation is the mass-conservation equation for the pollutant studied, i.e. eq.(2).

A *statistical model* is just based on regression functions resulting from a fitting of a set of observations. Its accuracy is obviously a function of the correlation coefficient between the observed data (here, the concentrations) and the quantities to which they are related by the regression function.

If the model is deterministic, it can be numerical or analytical. Strictly speaking, this terminology is inappropriate, since any type of model running on a digital computer is a sequence of operations on numbers, and is thus numerical. But this terminology refers to the mode of resolution of the differential equations.

Analytical models make use of an analytical solution of the differential equations, where the concentration to be calculated is an analytic function of input variables. In *numerical models*, the differential equations are replaced by a set of algebraic equations (e.g. finite difference equations) solved numerically.

The distinction between eulerian and lagrangian is based on the nature of the coordinate system used. *Eulerian models* make use of *fixed* coordinates. A spatial grid (see Fig. 10) is defined, and the concentrations are calculated at the grid points.

If the model is *lagrangian* the coordinate system is *mobile* and "follows" the wind trajectory.

An important category of analytical models are the gaussian-plume models. The basic formula is:

$$\chi = \frac{S}{2\pi \sigma_y \sigma_z \bar{u}} \exp\left[-\frac{y^2}{2\sigma_y^2} - \frac{(z-H_e)^2}{2\sigma_z^2} \right] \qquad (16)$$

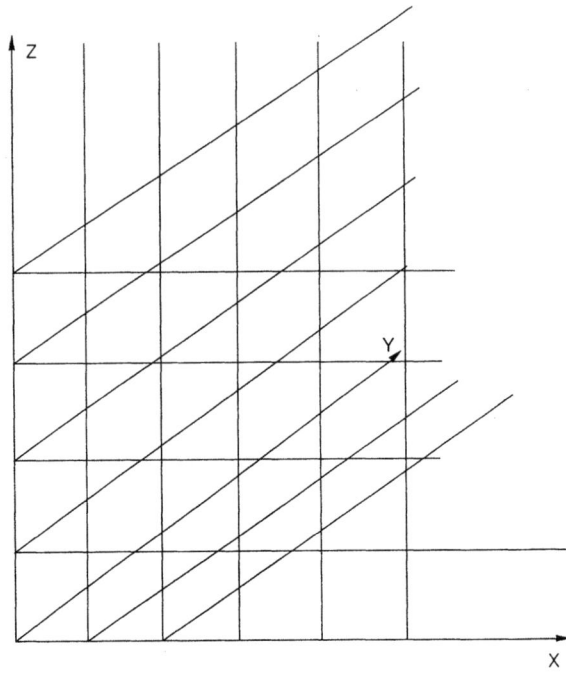

Fig. 10a - In a eulerian model, the coordinate system is fixed to the solid earth and appears in the form of a x-y-z grid.

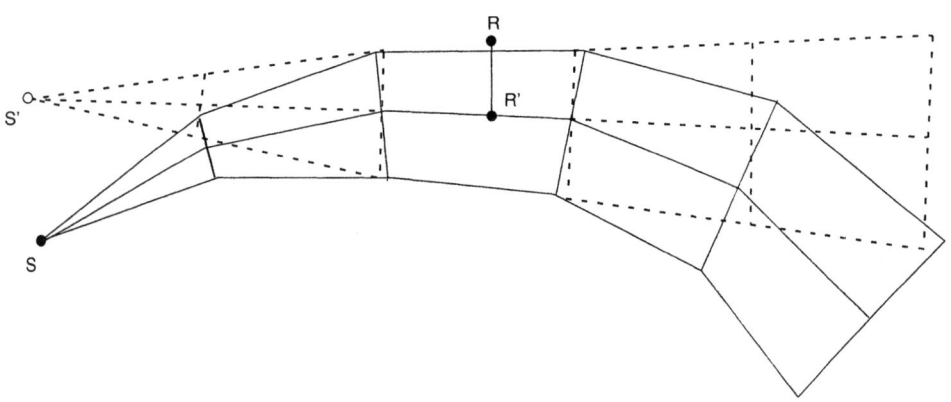

Fig. 10b - Typical scheme for a lagrangian model, where the coordinate system "moves" with the volume of air that is transported by the wind.

where:
- x = x in (x,y,z) downwind to the source in (0,0,H)
- H_e = H + ΔH (ΔH plume rise)
- S = emission rate
- ū = wind speed (along x)
- σ_y, σ_z = horizontal and vertical standard deviations, function of downwind distance x.

This formula is the analytical solution of eq.(2) in the case of steady-state and flat terrain conditions. Gaussian models are widely used because of their numerical simplicity. They are well adapted to multiple-source problems. The basic formula (16) is often modified in order to take more complex effects into account.

Fig. 11 shows a picture of the gaussian diffusion; the reason of calling it gaussian appears clearly, the concentration profiles along the y and z axes being gaussian functions.

The values of the standard deviations, σ_y and σ_z, of these gaussian functions (they express the "broadness" of the plume) depend on atmospheric stability.

We see now the fundamental role of that concept in the dispersion of pollutants. The dependence between the 's and the various expressions of stability can be found in the literature. The most classical is the relation with the Pasquill stability classes, which is shown on Fig. 12.

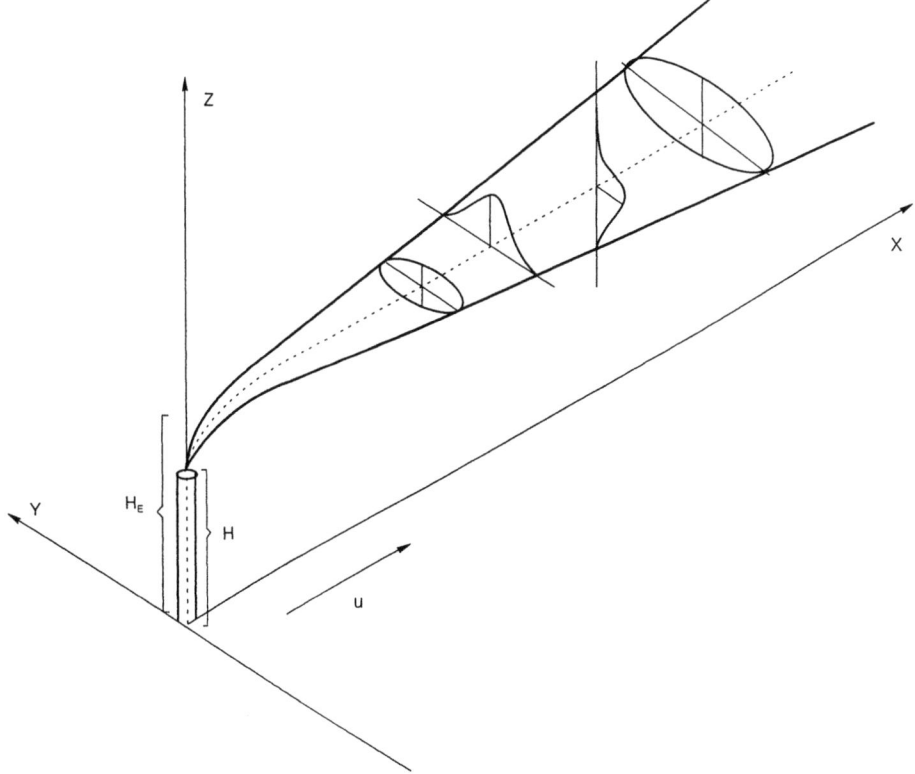

Fig. 11 - Visual representation of an idealized bigaussian plume, showing that the vertical and transversal concentration profiles across the plume are gaussian functions.

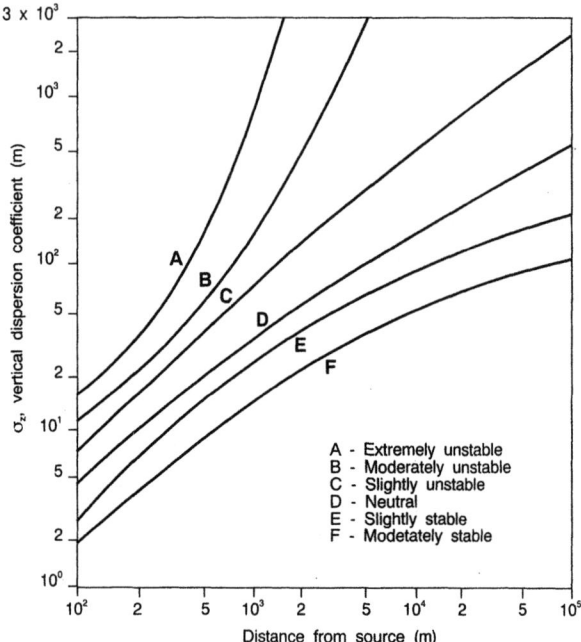

Fig. 12a - Relationship between σ_y (lateral dispersion) and stability, versus distance from source.

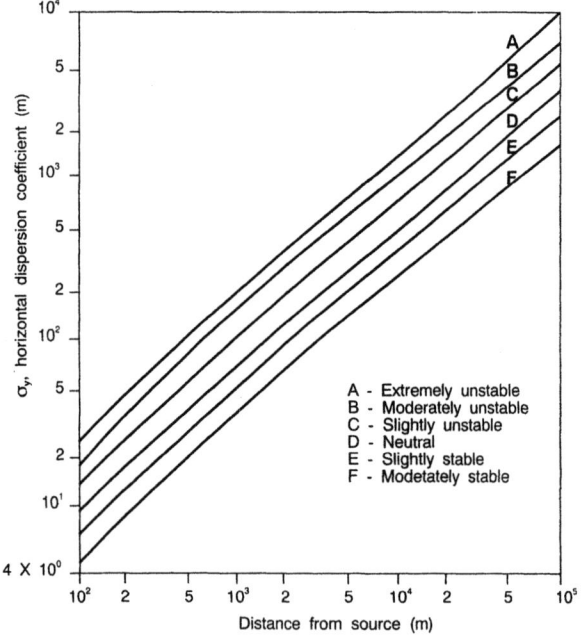

Fig. 12b - Relationship between σ_z (lateral dispersion) and stability, versus distance from source.

4.3 METEOROLOGICAL OBSERVATIONS

To run a model, we need input data. These are mainly of meteorological nature, as appears from the previous chapter. A good picture of the three-dimensional wind field is necessary, as well as temperature and radiation data.
Meteorological information are available from:
- instrumented towers (with cup anemometers and thermometers);
- pilot balloons and radiosondes;
- captive balloons;
- tracking tetroons;
- instrumented aircraft;
- doppler radars;
- acoustic sounders (Sodar and Doppler Sodar);
- satellites;

and from weather forecast models.

Frequently the determination of the wind speed and direction is far from being easy. The wind may vary in a quite complex way with height, with horizontal position (especially on complex terrain) and with time. These situations are usually the light wind situations and the plume will not travel very far in a few hours. Concentrations will tend to be relatively high, partly because of the low wind speed and partially because vertical mixing is very limited in these situations. This is particularly true at night. The dangers to local environment are therefore at maximum. More normally the wind speed is high enough for a reasonable degree of synoptic control; that is, the movement of the plume is governed by winds determined implicitly by the larger scale weather patterns and their associated pressure gradients. This does not mean that the underlying surface ceases to be important; on the contrary it remains of considerable importance in the lower atmosphere, but on a much broader scale so that every tree, house and hill, only has a relatively small influence in the integrated effect of the surface on the winds carrying the plume.

A standard procedure is then followed to obtain quantitative estimates of the effects of atmospheric stability, including thermal convection, on the dispersion. The mixing height may also be estimated. The procedure involves using latitude, longitude, time of day and year to determine the solar incident radiation on the surface or the top of the cloud layer and an estimate of the surface roughness and the thermal characteristics of the site. The Pasquill scheme, already mentioned, is one of the best known. Such models are appropriate for application to low level releases from single plants and may also be used for application to multiple source releases in areas of few hundreds of km^2, provided that the wind field is homogeneous, i.e. complications due to topography and sea- or valleys breezes are absent.

4.4 TRACER RELEASES

The tracer technique is the only one that permits to really follow a plume through the atmosphere. Its disadvantage, however, is its high cost, that prevents a frequent use.

An atmospheric tracer is a gaseous substance that does not "exist" in the atmosphere, and that is released in well-defined conditions. The "map" of its dispersion through the atmosphere is subsequently obtained by sampling (e.g. in plastic bags filled by pumping the surrounding air) followed by chemical analysis.

The mostly used tracers are sulphur hexafluoride (SF_6) and perfluorocarbons (C_7F_{14}; C_8F_{16}). All of them are nontoxic, nondepositing and chemically inert. They have a very low detection limit (1 ppt for C_7F_{14}).

5. Conclusions

In this chapter, we outlined the fundamentals of the processes involved in pollutant dispersal and transport through the atmosphere.

Emphasis is given on the role of meteorology, that governs the phenomenon; the methods of assessing air pollution are briefly described: modelling, measurements, tracer releases.

The knowledge gained during the last four decades in this field permitted to reach practical conclusions about problems of regional planning: it has, for example, become possible to minimize pollutant concentrations by placing the industrial sources at certain locations instead of others. Similarly, by choosing properly the height of the stacks the concentration at ground level can be reduced. This is generally obtained by constructing very tall stacks.

These methods, however, do not really solve the problem of air pollution, because the *quantity* of pollutants injected in the atmosphere remains unchanged. Further, by using high stacks, the probability of penetration of the pollutants into the higher atmospheric layers (the free atmosphere) is enhanced. When the pollutants are in the free atmosphere, they are entrained in the synoptic flow and can be transported over very long distances (up to 1000-2000 km); they remain thus for long times in the air and can than be subject to chemical transformations, giving rise to the acid rain problem (which is beyond the scope of this paper). So, by solving a problem (i.e. by developing techniques of reducing concentrations based on the knowledge of dispersion processes), one created unwittingly another problem.

The only real way of reducing air pollution resides in emission abatement policies: this consists of various techniques like putting absorbent filters in stacks, using catalysts on cars and fuels that have been previously desulfurised, and so on.

This does not mean that the knowledge of the mechanisms of dispersion and transport is not useful, but this is not the only solution.

The last word will then be a truism: the only way of reducing air pollution is to pollute less.

REFERENCES

Gaffney, J.S., Streit, G.E., Spall, W.D. and Hall, J.H., "Beyond acid rain", Environ. Sci. Technol., **21**, 519 (1987).

Hosker, R.P. and Lindberg, S.E., "Review: atmospheric deposition and plant assimilation of gases and particles", Atmos. Environ., **16**, 889 (1982).

Pasquill, F., "The estimation of the dispersion of windborne material", Met. Mag., **90**, 33 (1961).

Pasquill, F., "Atmospheric diffusion", J. Wiley, New York, (1974).

Paulson, C.A., "The mathematical representation of wind speed and temperature profiles in the unstable atmospheric surface layer", J. Appl. Meteorol., **9**, 857, (1970).

Pielke, R.A., "Mesoscale meteorological modeling", Acad. Press, Orlando, (1984).

Sandroni, S. amd Cerutti, C., Evaluation of air quality in an urban area", Toxicol. Environ. Chem., **20-21**, 11 (1989).

Stull. R.B., "An introduction to boundary layer meteorology", Kluwer, Dordrecht, (1988).

Swinney, H.L., and Gollub, J.P., "The transition to turbulence", Physics today, p. 41, August 1978.

Turner, B.D., "Workbook of atmospheric dispersion estimates", US Public Health Serv. Publ. 999-AP-26, pp. 1-84 (1969).

CONTROL OF AIR POLLUTION

The European Community Approach

H.P. Stief-Tauch
Head of Unit "Emissions from industrial installations and products"

1. **INTRODUCTION**

    Community Environment Policy was instituted by the Heads of States and Governments at their Summit meeting in October 1972 in Paris.

    Following this political initiative, a first Community Environmental Action Programme was adopted by the Council of Ministers in late 1973 laying down the principles for Community activities in this field. It contained a long list of specific measures to be taken as soon as possible in order to tackle the urgent environment problems occurring in the Member States. Principle lines of action were identified in the fields of water pollution, nature protection, international cooperation and, to a lesser extent, in product standards for tackling air pollution, particularly from SO_2 emissions.

2. **THE LEGAL SITUATION**

    The original Treaty of Rome, signed in March 1957, foresaw nothing on the environment, so that the legal basis for Community actions in this field could be built only on the "omnibus" Article 235. This Article serves as a basis for necessary actions not foreseen in the Treaty, but requires that the decision of the Council of Ministers is taken by unanimity.
    Above all, this is the main reason why, in the early years of EC Environment Policy, the achievements were not really as good as they might have been ("principle of least common denominator").

    The revision of the Treaty of Rome by the "Single European Act", which came into force in July 1987, has changed this situation : a new title VII "Environment" has been introduced which defines the EC's task in this field by giving, in the Articles 130 R to 130 T, detailed prescriptions of the principles and procedures to be followed. Amongst these are in particular :

    - the preference for preventive action,
    - 'the polluter pays' principle,
    - the integration of the environmental dimension into other Community policies,
    - the choice of the appropriate level of action : EC, national, regional (sometimes known as "subsidiarity"),
    - the possibility to take legal decisions not only unanimously but also by a qualified majority.

In cases where environmental matters are linked with actions aimed mainly at the completion of the internal market, a new Article 100 A of the Treaty allows the Council to decide by a qualified majority, with an increased possibility for the European Parliament to take a greater part in the decision process. Proposals made by the Commission on this legal basis must take as a basis a high level of environmental protection.

In this new legal context, the Commission has prepared and presented to the Council its Fourth Action Programme on the environment (1987-1992) and is currently implementing its contents.

3. **THE POLICY APPROACH - GENERAL CONSIDERATIONS**

In the following I will focus on the air pollution part for which I was and still am responsible to a large extent and

- give a review of the main lines of actions persued;
- explain the objectives, the basic considerations and the solutions retained;
- indicate major problems encountered and
- give an outlook on further improvements required or desired.

A comprehensive policy in the field of pollution control should address in a logical way major environmental problems. Consequently, the Community action has developped along three major lines, i.e.

a) by the <u>setting up of standards</u> in the form of

- air quality standards and objectives
- source performance standards (mobile and stationary ones). (The specific objectives of these standards, their potential and limitations will be discussed later on).
- product standards

b) by <u>monitoring the state of the environment</u>

c) by <u>promoting research</u> aimed at a better understanding of the phenomena involved.

The business of standard setting is certainly of particular interest to our lecture of to-day, the way how it was done, the solutions retained, the problems encountered and further improvements required/desired.

4. **ACTIONS IN THE FIELD OF AIR QUALITY**

Up to date, the Council has adopted three directives which lay down air quality standards ; they are the following :

- Council Directive 80/779/EEC of 15 July 1980 on air quality limit values and guide values for sulphur dioxide and suspended particulates, recently modified by Council Directive 89/427/EEC of 21 June 1989.

- Council Directive 82/884/EEC of 3 December 1982 on a limit value for lead in the air.

- Council Directive 85/203/EEC of 7 March 1985 on air quality standards for nitrogen dioxide.

Objectives

All these directives, established on grounds of urgency of the problems, aim first at the protection of public health and have for second aim the protection of the environment by laying down limit values for the ambient air concentrations of these pollutants not to be exceeded anywhere in the EEC. Moreover, guide values are given to ensure further protection.

Basic considerations and solutions

The air quality standards retained are based on the work of WHO expert groups as updated by medical experts from the Member States. Given the legally binding nature of the limit values (standards) which can be claimed by all citizens before the Court of Justice, care has been taken to lay down in the directives reference methods for sampling and analysing the pollutants concerned, as well as the obligation to monitor regularly the air quality where there is a likelyhood that the standards are being approached or exceeded.

For practical purposes, the Member States are offered the possibility to continue the use of their own national methods provided they have demonstrated to the Commission beforehand that the results of their methods are comparable to those obtained with the reference method. The reasons for such an approach are obvious : impossibility to scrap the existing networks (finance), long standing habits of national institutions to use certain methods rather than others, need for comparable results (intercomparison programmes required).

There are many places within the Member States where standards (limit values) are still being exceeded, mainly in highly industrialized areas. These areas require a substantial modification of the emission pattern from the relevant source categories, which would eventually mean major investment programmes. Such programmes must be defined properly, after an assessment of the local/regional situation and remedial measures must be implemented. The directives foresee for these "black" areas a certain delay within which the concentrations of the relevant pollutants must be brought down below the limit values. Cases where these values have been exceeded must be recorded and communicated to the Commission.

Member States are invited to lay down tighter national standards for areas which in their judgement require specific protection (e.g. national reserves). Up to now, only the Netherlands have used this possibility.

The SO_2 standards of Directive 80/779 take into account the effects on public health from the combined occurrence of SO_2 and suspended particulate matter (SPM). This directive sets also standards for SPM (dust) alone.

As far as the guide values are concerned, their legal nature is not a binding one. Numerically, they are substantially lower, i.e. more stringent than the limit values because of their double function as a long term goal for the air quality and as a reasonable level of protection for the living ecosystems. Indeed, man is rather a resistant species compared to many other living organisms and for a start, the EEC environment policy aimed at protecting man rather than ecosystems.

Problems encountered

It took me 4 years of tough negotiation in the Council to obtain acceptance by the Member States of the concept of a legally binding limit value (= air quality standard) for the Community as a whole, not to the least due to the lack of environmental consciousness at that time (the late 70's). The next two directives were passed in much shorter time, less than 2 years because the ground was broken and the need for EC legislation in this field became obvious. This is particularly relevant for the NO_2 directive 85/203/EEC which was rapidly adopted in a situation where the dieback of forests gave large concern to the public.

Another serious obstacle was the existence of a wide variety of measurement methods considering different forms/species of dust (SPM in some cases, but mostly black smoke - BS). Both concepts have been taken on board in Directive 80/779, but the price to be paid was a major effort for the Commission together with the Member States to run a large intercomparison measurement programme, followed by a quality assurance programme. Without the very active collaboration of our colleagues here in the JRC-Ispra, the Commission would never have been able to carry through this exercise with its valuable results and I wholeheartedly give them all the credit they deserve for the enthousiasme with which they have done this job (which, by the way, is still continuing).

These ongoing activities are an important contribution to the technical implementation of these directives within the Member States.

As an example I would mention :

- the quality assurance programmes for the SO_2/SPM directive which have largely improved the quality and comparability of the pollution measurements made within the Member States in the

- network stations and national reference laboratories (an important task, given the economic consequences of legally binding air quality standards);
- the studies on the <u>design of air pollution measurement networks</u> for SO_2, SPM, BS and NO_2 (financed by the Commission and carried out by various laboratories and European experts) ;
- the <u>evaluation of specific networks</u> as to ensure their conformity with the tasks required under the air quality directives (networks of Paris, Madrid, Portugal);
- the <u>organization of seminars and symposia</u> destined to enlarge experience between technicians and instrument suppliers.

Further improvements required/desired

These activities have, to a certain extent at least, contributed to a better implementation of the directives at national level, although there are still many shortcomings which give rise to legal action taken by the Commission against the Member States. In particular, the flow of information on the pollution levels measured (or rather not measured) communicated to the Commission is not as regular as it should be.
Therefore the population has only an insufficient awareness of the quality of the air they breathe. To remedy this situation, the Commission decided recently to allow public access to its data base containing the information on national legislation taken in application of Community law.

At the last Council meeting of Environment Ministers on 7 June 1990, an important Directive 90/313/EEC on public access to environmental information has been adopted, which will further the transparency of knowledge on the state of the environment within the Community.

This directive will correct the deficiencies of many EEC and national legal provisions which had not given enough attention to the divulgation of environmental data.

The creation this year of a <u>European Environment Agency</u> also seeks to improve the information deficit. Its main task, at least in the beginning, will be the collection of data on the state of the environment. In the field of air pollution (control) this will involve the establishment and updating of more complete emission inventories for the main pollutants, emitting sources and source categories – a task already started some years ago in the frame of the CORINE project of the Community. (CORINE = Coordination, Information, Environment).

A much more representative data base for the effectively measured air quality in all Member States is to be established as well because the 3 air quality directives only require Member States to report cases where the limit values have been exceeded.

It is only in the course of extensive research carried out within the EEC and outside, partly in the frame of the COST programmes, that one realized the full extent of the complicated interactions between the cocktail of pollutants from the air or the water, and living ecosystems. The scientific conclusions one can draw to-day indicate a clear need for much tougher quality standards (air and water) than the old guide values if policy is aiming at an effective protection of sensitive ecosystems. A new receptor-oriented approach is currently being forwarded by the international scientific community called the <u>critical load concept</u>. This concept appears as a logical improvement of the EEC environment policy although its implementaion will cause many problems and will require difficult political choices not to the last on the extent of protection which the economies can pay.

A <u>directive proposal</u> is currently under preparation on the laying down of <u>air quality objectives for ozone</u>. This is a new approach towards controlling a so-called secondary pollutant, formed by the interaction of NOx and volatile organic compounds (VOC) in the presence of sunlight. Contrary to mandatory air quality standards, air quality objectives have an indicative character, so they may be exceeded. The directive will require Member States to monitor ozone and its precursors, compare the levels recorded with the objectives set, to elaborate strategies and take measures for the reduction of precursor emissions from relevant sources in order to bring gradually the ozone levels down below the quality objectives.

5. ACTIONS IN THE FIELD OF EMISSION STANDARDS

Air pollution is moving as the wind blows, without respecting any boarder, sometimes over thousands of km.
The pollutants undergo thereby many physico-chemical modifications, they are deposited and the acidic deposits of sulphur and nitrogen emissions give rise to special concern (the famous acid rain).

A policy looking at the air quality only is of little help against this phenomenon : what is required rather is a policy controlling the emissions, and preferably at the source already.

In principle two main routes are open at this end :

- to act on the technical performances of the emitting source, eg. by improving the efficiency of combustion processes or by treating the combustion gases;

- to act on the composition of fuels and by setting requirements concerning the use of certain fuels in specific installations only (eg. by limiting the sulphur content in gas-oil which is widely used for domestic heating purposes and for transportation).

In some cases both routes must be taken simultaneously to achieve the desired reduction rates, as it was the case in the EEC environment legislation where product standards and source performance standards were established together.

5.1. Mobile source emission standards

The first standards in this field were set as early as 1970 limiting the emissions of CO and unburnt hydrocarbons (HC) from the tailpipe of motor vehicles (passanger cars). The motivation at that time was not environment protection but elimination of technical barriers to trade. Uniform prescriptions were required for the composition and properties of goods widely traded between the Member States which were/and still are very ingenious in inventing all sorts of special requirements aimed at the protection of their domestic markets and their industries.

(In this context one should not forget that also environmental requirements can serve for this purpose, thus the Community has an important task to fulfill by harmonizing as far as politically and economically feasible/acceptable these requirements).

Up to the mid 80's these standards were successively strengthened as technology progressed, the NOx emisisons were regulated as well and the recycling of crank case gases required;

While discussing in the early 80's at the Council of Ministers the proposed quality directive on lead in air, it became evident that the main source was the leaded petrol. To achieve the required standards there was only one practical way : to lower as soon as possible the lead content and to bring unleaded petrol on the market; it would have been unrealistic to count upon a decrease in trafic volume when the fleet of vehicles on the road increased steadily in all Member States.

By introducing unleaded petrol another technical prerequisit for the further tightening of emission standards for motor vehicles with gasoline engines was fulfilled : catalytic devices could then be applied allowing a drastic reduction of all three tailpipe emissions (CO, HC and NOx).

Up to the standards set by Directive 83/351/EEC, the requirements for controlling motor vehicle emissions were relatively easy to fulfill by using classical control devices (well adjusted carburettor, fuel injection, precise ignition timing). With rising environmental concern of the population however, particularly on the NOx emissions (where trafic is in some countries the main source – see maps in annex), pressure grew on the national legislators and on the

Council of Environment Ministers to strengthen substantially the requirements on controlling these emissions, the more as appropriate control technology did exist and was mandatorily required for all cars destined for the US market. The EEC manufacturers exported their products to a variable extent to this market and were not too enthousiastic about having the same tough requirements being imposed on their entire domestic production (problems invoked were, among others : lack of production capacity, unsupportable cost increase for the consumer, unavailability of high quality unleaded petrol, etc.).

The legal basis for EEC legislation being Article 100 requiring decisions by unanimity, progress was very slow before the entry into force of the revised Treaty of Rome (July 1987). A look on the main lines of development during these years is quite interesting :

- In June 1985, the Council defined in its "Luxembourg compromise" the great orientations concerning the calendar and the level of emission standards for the passenger cars with big engines (above 2 litres) and medium sized engines (1,4 up to 2 litres);

- Beginning 1987, still no formal juridical decision on this orientation in the form of a new directive because 3 Member States disagreed (DK, GR, NL).

- With the entry into force of the Single European Act in July 1987, the Commission presented a revised proposal based now on Article 100 A (requiring only qualified majority) and the Council finally adopted in December 1987 the Directive 88/76/EEC giving legal force to the Luxembourg compromise.

  It should be recalled that for cars with big engines this directive practically obliges the manufacturer from October 1989 onwards, to use the 3-way catalytic devices with closed loop control to fulfill the standards (like in the US – see annex tables 1 & 2), while cars with medium and small engines are able to fulfill the standards with a much less sophisticated control technology.
  Standards for small engines below 1.4 L were fixed as a first step only, a new proposal from the Commission was expected to lay down the definite standards before 1989.

Directive 88/76/EEC contains also some important supplementary elements :

- the standards laid down for gasoline engined cars apply, in principle, with some modulations, to diesel engined cars as well;
- all new motor vehicles with gasoline engines falling under this directive must be able to operate with unleaded petrol;
- the test procedure for the type-testing of new vehicles will be revised to incorporate also extra-urban driving conditions;
- the first emission limits for particles from passenger cars and light duty vehicles driven by a diesel engine were laid down in Council Directive 88/436/EEC of 16 June 1988 to become effective from 1 October 1989 onwards.

Directive 88/77/EEC was adopted by the Council on 3 December 1987 limiting for the first time the emissions of gaseous pollutants from the tailpipe of diesel engines in lorries and buses (above 3.5 tonnes), moreover requiring the Commission to come up quickly with new proposals for further strengthening of the EEC legislation for this source category.

With the next Directive 89/458/EEC adopted in June 1989 establishing the second stage of emission standards for passenger cars with small engines (below 1.4 L), the approach of the EEC towards controlling mobile sources changed dramatically. For the first time the Commission managed, with the active support of the European Parliament, to convince the Council of Ministers that the time had come to require also this industrial sector to fully apply the existing technological means for the benefit of an efficient pollution control of motor vehicles, giving European citizens the same degree of protection as American or Japanese ones.

The new standards promulgated can only be achieved if the currently "best available technology not entailing excessive cost", i.e. 3-way catalytic devices with closed loop control is being used, in particular with the new enlarged European test cycle with its additional non-urban driving sequence.

The concept of requiring the use of best available technology not entailing excessive costs upon licensing new stationary industrial sources has been introduced into EEC environment legislation since 1984 with Directive 84/360/EEC, but was not accepted for controlling mobile source pollution before 1989.

Two further directive proposals are still pending before the Council, the first one is extending this concept uniformly on medium and large sized passenger cars with either petrol or diesel engines, requiring moreover a control of evaporative losses of hydrocarbons, based on the enlarged European test cycle and being mandatory from 1 July 1992 on for new models, from 1 January 1993 for all new vehicles.

The other proposal strengthens further in 2 stages the emission standards for diesel engined heavy vehicles (above 3.5 tonnes), for gaseous pollutants and for particulates.
All these efforts deployed over the last four years towards tighter emission control of mobile sources are part of the overall policy approach to stabilize and reduce air pollution, in particular NOx where the transport sector constitutes already the major source category in a number of Member States. (See maps in annex).

5.2. **Stationary source emissions standards**

Major sources of NOx (and above all of SO_2) are stationary combustion installations and processes which have been at the focus of Community legislation since the early 80's.

Up to now five directives on industrial emissions have been adopted by the Council, laying down general obligations and specific requirements for the operation of certain plants including emission standards ; they are the following :

- Council Directive 84/360/EEC of 18 June 1984 on the combatting of air pollution from industrial plants;

- Council Directive 87/217/EEC of 19 March 1987 on the prevention and reduction of environmental pollution by asbestos;
- Council Directive 88/609/EEC of 24 November 1988 on the limitation of emissions of certain pollutants into the air from large combustion plants;
- Council Directive 89/369/EEC of 8 June 1989 on the prevention of air pollution from new municipal waste incineration plants;
- Council Directive 89/429/EEC of 21 June 1989 on the reduction of air pollution from existing municipal waste incineration plants.

Objectives

Directive 84/360/EEC sets general requirements imposed upon new industrial plants and existing ones (obligation of a licence; conditions under which a licence can be given to prevent air pollution; exchange of information between Member States and the Commission on experience achieved on prevention and reduction measures, on technologies and equipment used, on emissions and air quality limit values; follow-up of the evolution of best available technology not entailing excessive costs (BATNEEC) and its application to existing plants - retrofit). It also allows the Commission to elaborate specific proposals for Community wide emission standards.

The other 4 directives are specific ones, dealing with a particular sector of industrial activity or a specific pollutant (in the case of asbestos where not only emissions into the air have been addressed but also into water as well as the safe disposal of asbestos containing wastes (integrated approach). They set emission standards to be respected for new plants throughout the EEC (87/217/EEC, 88/609/EEC, 89/369/EEC) and also for the adaptation of existing plants (89/429/EEC) or for the overall decrease of emissions from all existing plants of this category by imposing a "bubble" on national emissions (88/609/EEC).

In all cases specific requirements for regular monitoring of the emissions with agreed methods are foreseen, and reporting obligations are established.

Problems encountered

It would fill a whole day's lecture if I were to enter into the details of each directive and I must limit myself to some major problems which had to be addressed by us in the course of the preparation and negotiation of the two most important directives : the general one (84/360/EEC) and the first specific one (88/609/EEC).

Whilst proceeding, I would mention that both the first directive on air quality standards for SO_2 and SPM (80/779/EEC) and the first directive establishing emission standards for a category of major stationary sources, the large combustion plants (88/609/EEC) are the object of intense studies in universities for doctorate thesis, in order to analyse all sides of the genesis of these basic pieces of Community environmental policy.

Directive 84/360/EEC

Certain categories of industrial plants listed in Annex I of the Directive are subject to prior authorisation. They were chosen on grounds of the nature and magnitude of their emissions (SO_2, NOx, HC, heavy metals, dust and fibres, halogenes - energy industry, production and processing of metals, manufacture of non-metallic mineral products, chemical industry, waste disposal, paper pulp).

Major modifications of such existing plants also require an authorisation, and, in all cases, national competent authorities are responsible for this job. For a number of Member States, the list was too short - they can extend it if necessary; for others, it was too long - they will have to adapt their national practice.

The main requirement (Article 4) for the release of an authorisation by the competent national authority is that

1. "all appropriate preventive measures against air pollution have been taken, including the application of the best available technology, provided that the application of such measures does not entail excessive costs", (BATNEEC);

2. "the use of a plant will not cause significant air pollution particularly from the emission of substances referred to in Annex II;

3. none of the emission limit values applicable will be exceeded;

4. all the air quality limit values applicable will be taken into account.

You will not find in the Directive a precise definition of what is meant by BATNEEC; this is left to the discretion of the national authorities.
The formulation of the prinicple has been kept vague on purpose in order to give a margin of appreciation which is necessary in a Community with widely differing situations (in legislative, structural and economical terms).

One may ask - how is the aspect of a gradual harmonization of environmental protection measures being dealt with, even more so as also the question of conditions of competition is at stake ?
The answer is twofold : on the one hand, Member States are obliged to exchange information among themselves and with the Commission on the application of this Directive (Article 7), an exercise which is well under way; on the other hand "they shall follow developments as regards the best available technology (no notion of the cost element here!) and the environmental situation" (Article 12).
On top of this, there is the possibility offered in Article 8 :

   1. The Council, acting unanimously on a proposal from the Commission, shall if necessary fix emission limit values based on the best available technology not entailing excessive costs, and taking into account the nature, quantities and harmfulness of the emissions concerned.

   2. The Council, acting unanimously on a proposal from the Commission, shall stipulate suitable measurement and assessment techniques and methods.

This same Article is the basis for all the specific directives elaborated since and for which the Commission, in close collaboration with national experts and the industries concerned, laid down what is the current "state of the art" (BATNEEC). The requirement of a Council decision taken by unanimity, implicates of course the danger of a "least common denominator" legislation not reflecting the true technological capability of the industry (remember the struggle on motor vehicle emission standards as long as the unanimous decision was required and the modest results achieved at that time).

The most difficult feature of this directive is the retrofit requirement for existing plants of the categories listed in Annex I. Such an obligation was new for the legal systems in most Member States where an authorisation, once it was delivered, remained in force as long as the plant was operating without substantial modifications.

Article 13

In the light of an examination of developments as regards the best available technology and the environmental situation, the Member States shall implement policies and strategies, including appropriate measures, for the gradual adaptation of existing plants belonging to the categories given in Annex I to the best available technology, taking into account in particular :

- the plant's technical characteristics,
- its rate of utilization and length of its remaining life,
- the nature and volume of polluting emissions from it,
- the desirability of not entailing excessive costs for the plant concerned, having regard in particular to the economic situation of undertakings belonging to the category in question.

The term "best available technology" is in this specific case subject to a number of constraints pertaining to the individual situation of the installation concerned, hence the resulting requirements being imposed upon existing plants will be generally less stringent than for new installations.

Further improvements required/desired

The obligation of using BATNEEC as a basis for authorizing a new plant requires an important effort from competent national authorities, particularly in those Member States where national legislation in this field is less developped.

The very general definition given by the Directive which covers a large number of heterogenous sectors and an even larger number of processes, requires the establishment of a procedure to identify and follow-up the technologies in question.
The Commission has enquired into the measures taken by the Member States in application of Articles 7 and 12.
The main conclusion one could draw was that considerable difficulties do exist for an efficient and harmonized application

of these provisions. For this reason, the Commission considered that the exchange of information foreseen by Article 7 had to be organized and structured with a view to <u>prepare technical notes defining the best available technologies</u>, the conditions of its application, the resulting emission levels and other important aspects for the implementation of the Directive.
The structure of these technical notes can be summarized as follows :

- definition of the category of industrial installations dealt with in the technical note;
- description of the available industrial processes;
- study of measures available for pollution reduction;
- study of multi-media aspects;
- designation of the best available technologies for reducing the pollution and the resulting emission levels of different pollutants for each identified technology. These emission levels are expressed in the form of emission limit values to the air for the individual pollutants. These levels will serve as criteria for the choice of technologgy and for the authorisation procedure of the industrial installation concerned;

- study of methods for controlling the pollutant emissions;
- study of procedures for storage of base materials or products;
- study of measures required to minimize the emissions during the start-up of the installation.

For the time being, the activity of drafting and agreeing on the technical notes is structured as follows :

- technical working groups consisting of representatives of the Commission, Member States and of the industrial sectors concerned have been set up;
- these working groups draft the technical notes and submit these for approval to a coordination committee consisting of the same categories of participants. This committee has also the task of defining priorities for the drafting of technical notes, in particular the choice of priority industrial sectors.

During the pilot exercise of 1989, the following technical notes have been established and approved :

- production of sulphuric acid;
- production of nitric acid;
- production of ammonia;
- fabrication and storage of benzene;
- production of cement;
- incineration of toxic and dangerous waste.

These technical notes will be available in English by the end of 1990 in the form of EUR publications.

For the period 1990-1991, the following technical notes are programmed :

- refining industries;
- cokeries;
- agglomeration of minerals, oxygen steel furnaces;
- electric arc steel furnaces;
- metal foundries and rolling wills.

A work of reflexion on the structure of technical notes and the future priorities in the field of controlling industrial emissions is under way at the Commission services with a view to facilitating and speeding up the implementation of the Directive.

A **new directive proposal** is currently under preparation with a view to establish Community emission standards for incineration plants of dangerous substances, and should be presented to the Council before the end of 1990 if possible.

Work is in hand on the **reduction of VOC emissions** from stationary sources in the frame of action taken to combat photochemical air pollution. This concerns the elaboration of draft directives aimed at a limitation of VOC emissions from **printing**, **metal degreasing**, **automobile painting** and the **private use of paints**.
The sectors of **dry cleaning**, **treatment of wood and of metallic surfaces** will be considered in a second phase (1991).

Finally, in 1990-1991 preparatory work will start to develop new directives eventually controlling emissions from

- gas turbines;
- combustion installations of less than 50 MW thermal power.
- combustion plants burning solid fuels of 50 to 100 MW thermal power.

Directive 88/609/EEC (large combustion plants)

It constitutes the introduction into Community policy in the field of air pollution control of the principle of EEC-wide emission standards for stationary sources and the first application of the concept of a national emission ceiling for emissions of existing plant (a "bubble") which is gradually lowered within an agreed calendar.

The main features are :

- all large combustion plants aboce 50 MW thermal power are covered;
- emission limit values (standards) are laid down for new or modified plant with respect to SO_2, NO_2 and dust;
- an emission ceiling for SO_2 and NOx from all existing plants is imposed on each Member State which is to be gradually lowered further within an agreed calendar:
- extensive measurement and reporting requirements are specified as well as the use of agreed equipment.

Problems encountered

A difficult matter was the <u>definition of best available technology not entailing excessive cost</u> as a basis of the emission standards. In preparing the proposal in 1983, I took the view that for the immediate future, only such technologies having given proof of their industrial viability could serve as a basis. This means e.g. for SO_2 abatement : choice of appropriate fuels, measures influencing the combustion process and FGD (flue gas desulphurisation), for NOx abatement : measures improving the combustion process (low NOx burners, staged combustion) but no catalytic treatment of the flue gases. The latter processes were just being introduced in the FRG at large scale but were full of uncertainties.

For a second stage of emission standards to be defined before 1994, the recent improvements in abatement technology will of course be taken on board.

The greatest political difficulty was the <u>determination of the reduction of emissions from existing plants</u>. From the start, it was obvious that a general requirement of retrofitting each plant would not find an agreement, therefore only the solution of a bubble approach at the scale of individual Member States appeared as a viable one.

(see maps in the annex on the importance of SO_2 and NOx emissions from this source category).

Nevertheless, the sharing out between Member States of the proposed EEC total reduction rates deemed necessary (-60% for SO_2 and -40% for NOx within 10 years - reference year 1980 - target date 1995) took 3 years at least because the principle of a uniform rate for each one was not acceptable. We had to incorporate the specific situation of each country in respect of its energy supply and future requirements, the nature of indigenous fuels used, social considerations and aspects of regional development, technological capabilities, not to forget the financial side of this operation (total cost estimated in 1985 : 35-50 billion ECU).

Many other problems of minor importance had to be settled as well, so it was no surprise to the Commission that it took 5 years from presentation of the proposal (December 1983) until the final adoption by the Council (November 1988) and the text of the directive is really the result of intensive horse trading between the Member States falling rather short of the initial proposal of the Commission.

6. **ACTIONS IN THE FIELD OF PRODUCT STANDARDS**

To combat air pollution from SO_2 and to improve in particular the air quality in urban areas, an action against low level emissions, mainly from domestic heating and small trade and industrial premises was felt necessary. The appropriate measure was the definition of a product standard for the sulphur content of gas-oil, widely used as fuel in these installations and in the transport sector.

Council Directive 75/716/EEC of 24 November 1975 on the approximation of the laws of the Member States relating to the sulphur content of certain liquid fuels, laid down uniform standards for 2 types of gas-oil (with 0.8 and 0.5% of S maximum).

In order to ensure free trading of these gas-oils throughout the Community, the legal basis chosen was Article 100 (requiring unanimity of the Council). The maximum allowable sulphur content was lowered further to 0.3% S by Council Directive 87/219/EEC of 30 March 1987, giving Member States the possibility to require, for specific zones with severe environmental problems from SO_2 pollution, the use of gas-oils with 0.2% S content.

Currently, the Comission is considering the need for the creation of a specific automotive gas-oil with extra-low sulphur content (0,05%) and a modified composition, allowing a further reduction of pollutant emissions from heavy diesel engines (above all NOx and particles).

Another product standard with bearing on industry was laid down by Council Directive 78/611/EEC of 25 June 1978 on the approximation of the laws of the Member States concerning the lead content of petrol. As for the gas-oil directive, the same trade considerations applied, again Article 100 was the legal basis.

The maximum permissible lead content was fixed in the range of 0.4 to 0.15 gr Pb/litre. When the need for the introduction of unleaded petrol arose the EEC legislation was modified by Council Directive 85/210/EEC of 20 March 1985 on the approximation of the laws of the Member States concerning the lead content of petrol.

Adopted in the wake of the oil crisis, this new directive required Member States to take measures to ensure the availibility and balanced distribution within their territories of unleaded petrol from 1 October 1989 (or earlier if they wished so).

The quality of this unleaded petrol (sometimes called "Euro-Super") is fixed as follows :

```
    Pb       < 0,013 gr/L
    benzene  < 5% volume (for all petrols)
    RON        95,0 min )
                        ) at the pump
    MON        85,0 min )
```

7. ACTIONS DEALING WITH GLOBAL ENVIRONMENTAL PROBLEMS

The overview of Community policy dealing with the control of air pollution would be onesided without at least mentioning shortly other legal measures adopted or in course of preparation. They constitute a response to problems, exceeding by far the frontiers of the EEC or even Europe.

Among these there is the 1979 Convention on long-range transboundary air pollution of the UN-ECE in Geneva to which the Community is Party, with 2 protocols on SO_2 and NOx emission reduction in force and another one on VOCs in hand. In the frame of this Convention, the European Monitoring and Evaluation Programme (EMEP), financed by more than 55% by the EEC and its Member States, constitutes the backbone of information on the quality of the atmosphere over Europe (one important measurement station is run by the Commission here in Ispra). The results of EMEP serve as a precious guide to define and implement policies aimed at combatting acid depositions within the European region.

An action of worldwide importance has been launched since 1978 concerning the chlorofluorocarbons (CFCs) and other substances threatening the stratospheric ozone layer. As Party to the Vienna Convention of 1988 on the protection of the ozone layer and to the Montreal Protocol established afterwards, the Community has taken a number of Council decisions and resolutions concerning this problem. The latest one of June 1990 sets the objective of eliminating by 1998 at the latest within the EEC all productions and uses of CFCs known to damage the stratospheric ozone layer (including certain uses of halons).

An even bigger problem which has already made a lot of headlines is the greenhouse effect on climate. Linked to the changing composition of our atmosphere attributed partly to human activities and the increasing consumption of fossil energy, the balance of important trace gases appears likely to disturb the future climate. CO_2, CH_4, N_2O, O_3 and other gases are emitted or formed at increasing rates, this may severely modify through climate changes our way of life and threaten basic resources of mankind (food, energy, raw materials).

In the frame of international actions (IPCC = Intergovernmental Panel on Climate Change, the World Climate Conference) the Community is actively engaged to elaborate solutions for its own territory and contributing to a concerted worldwide answer. Apart from more research (EPOCH programme for climate studies), a general reflexion on the ways how to address the problem and efforts to preserve and restore the tropical forests the main emphasis is put on sustained efforts for a much more rational use of energy in all major sectors of our economy which will bear automatically a benefit in terms of avoided emissions of air pollutants among which SO_2, NOx, CO_2, N_2O etc.

Two programmes have been launched this year at this end called THERMIE promoting the energy technologies and SAVE promoting higher energy efficiencies.

An important political goal is persued by the Community : to convince all major industrialized countries of the need

- to stabilize their CO_2 emissions by the year 2000 and to reduce them around 2010;
- to stop deforestation by the year 2000.

8. OUTLOOK ON FUTURE DEVELOPMENTS AND CONCLUSIONS

The lastly mentioned activity of finding appropriate answers at EEC and worldwide level to the potential threat of climate changes from greenhouse gases will by no doubt constitute our major task up to the end of this century. A big effort to save energy will help the EEC to bring further down the emissions of major air pollutants and provide better air quality.

International cooperation will more than hitherto be the keyword for tackling and solving major problems for the rehabilitation of environmental quality (Eastern Europe, the Third World).

The currently still prevailing policy of dealing with individual parts of the environment only (air, water, soil) has to be changed towards an integrated approach. My specific task in the field of industrial pollution control will be to mastermind a comprehensive policy approach, if possible, with an optimisation of the achievable environmental benefit from such an integrated approach.

The potential of using economic instruments as a complement to the legal big stick is currently being investigated and may yield cheaper and more rapid solutions in some cases.

You will note that 18 years of EEC environment policy have achieved a good start but the bigger part has still to come.

Category of vehicle (displacement volume)	Date of introduction of new emission standards		Emission standards to be respected (in grams per test)		
	New car models	New cars	Carbon monoxide	Total hydrocarbons and nitrogen oxides	Nitrogen oxides
More than 2000 cm³	1.10.88	1.10.89	25	6.5	3.5
Between 1400 and 2000 cm³	1.10.91	1.10.93	30	8	-
Direct injection (turbo) (diesel engines)	1.10.94	1.10.96	30	8	-
Less than 1400 cm³	1.10.90	1.10.91	45	15	6

TABLE 1
Standards for type approval of model or type of vehicle
(Directive 88/76/CEE of 3.12.87)

Category of vehicle (displacement volume)	Date of introduction of new emission standards		Emission standards to be respected (in grams per test)		
	New car models	New cars	Carbon monoxide	Total hydrocarbons and nitrogen oxides	Nitrogen oxides
More than 2000 cm³	1.10.88	1.10.89	30	8.1	4.4
Between 1400 and 2000 cm³	1.10.91	1.10.93	36	10	-
Direct injection (turbo) (diesel engines)	1.10.94	1.10.96	36	10	-
Less than 1400 cm³	1.10.90	1.10.91	54	19	7.5

TABLE 2
Standards of conformity of production
(Directive 88/76/CEE of 3.12.87)

Date of introduction of new standards		Standards for type approval			Standards for conformity of production		
New types of heavy vehicles (>3.5 t)	New vehicles (>3.5 t)	Carbon monoxide (g/kWh)	Hydro-carbons (g/kWh)	Nitrogen oxides (g/kWh)	Carbon monoxide (g/kWh)	Hydro-carbons (g/kWh)	Nitrogen oxides (g/kWh)
1.7.88	1.10.90	11.2	2.4	14.4	12.8	2.6	15.8

TABLE 3
Standards for the emission of polluting gases from lorries and buses (1st stage)
(Directive 88/77/CEE of 3.12.87)

Displacement volume	Date of introduction of new standards		Mass of particles emitted	
	New types of vehicles	New vehicles	Standards for type approval	Standards of quantity production
All	1.10.89	1.10.90	1.1 (g/test)	1.4 (g/test)

TABLE 4
Standards for the emission of particles from diesel engined cars (1st stage)
(Directive 88/436/CEE of 16.6.88)

DESULFURIZATION OF FLUE GASES ON THE BASIS OF LIME OR LIMESTONE SCRUBBING

G. MITTELBACH
Deutsche Babcock Anlagen AG
Postfach 4 + 6,
D-4150 KREFELD 11
Germany

1. General

The most widely applied technique for the desulphurization of flue gases is wet scrubbing where a slurry of water-gypsum-limestone or water-gypsum-lime is used as the scrubbing liquid.
This technique is based on developments over a long period in many industrial size plants. Fig. 1 shows a survey of the installed desulfurization capacities in Japan, USA and in the Federal Republic of Germany. It follows that by far in most processes limestone ($CaCO_3$) or lime (CaO) are applied as neutralization agents.

The installation of FGD systems was started in Japan at about 1970, shortly afterwards followed by the USA. In Europe, the first plants were erected in Germany nearly 10 years later. This time gap is also reflected in the distribution of the various applied desulfurization techniques in the different countries as shown in Fig. 1. A gradual selection between the available processes took place when the processes had to prove their technical and economical suitability on an industrial scale.

There are various reasons because the limestone process has become the most widely applied one for the desulfurization of boiler flue gases, i.e. for the treatment of large gas volumes with relatively low sulfur dioxide contents (in Germany mainly between 300 and 1000 vpm). The main reasons are the following:

- limestone is one of the cheapest chemical agents and also widely available as a natural product in most countries;

- the proces is simple and no "chemical factory". It can easily be adapted to load variations of power plant boilers;

- the by-product gypsum can be used in large quantities in other industries. However, it is also possible to dump the product, if this appears to be more convenient. Dumping causes no great trouble, because also gypsum is a natural product.

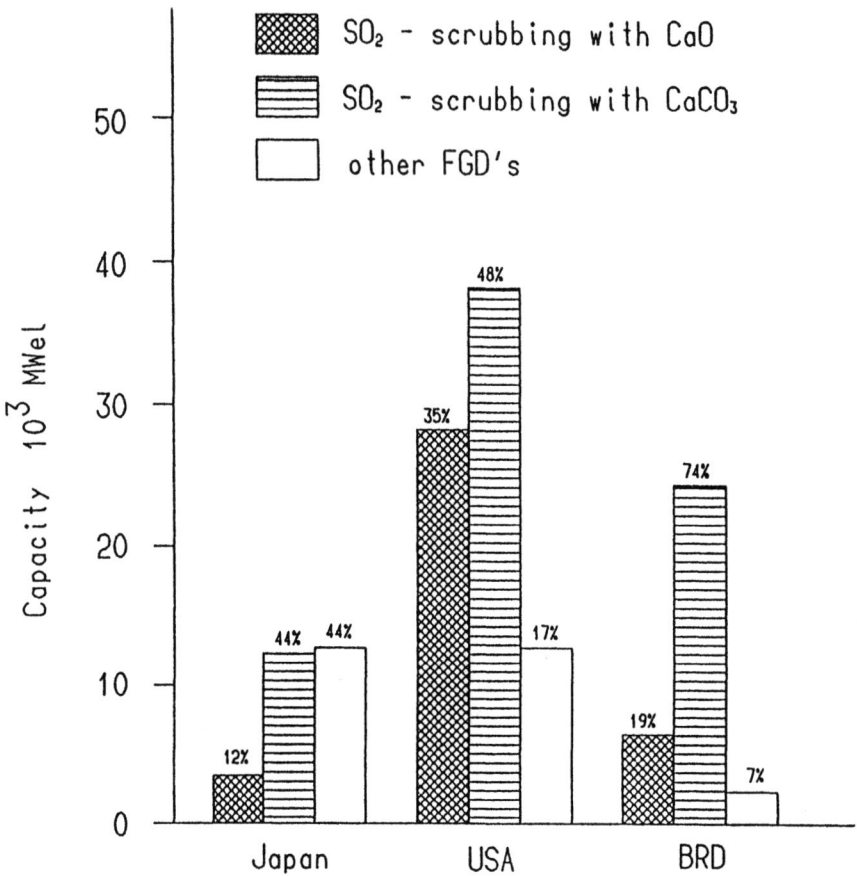

Fig. 1
Installed FGD capacities

2. Process principles

2.1 Process development
The overall reactions between $CaCO_3$ and the noxious components SO_2 (sulfur dioxide), HCl (hydrogen chloride) and HF (hydrogen fluoride) present in trace concentrations in the flue gas, are the following:

$$SO_2 + \tfrac{1}{2}O_2 + H_2O + CaCO_{3\,s} \rightarrow CaSO_4.2H_2O_{\,s} + CO_{2\,g}$$

$$2\,HCl + CaCO_{3\,s} \rightarrow CaCl_{2\,sl} + H_2O + CO_{2\,g}$$

$$2\,HF + CaCO_{3\,s} \rightarrow CaF_{2\,sl} + H_2O + CO_{2\,g}$$

There are some similarities between the limestone and the lime based desulfurization processes because in both cases:

- a solid of low solubility must be dissolved in the scrubbing liquid to neutralize the absorbed pollutant gases;

- the product of the reaction gypsum ($CaSO_4.2H_2O$) is a solid, which has continuously to be separated from the scrubbing liquid.

The main difference between the neutralization agents is the much higher solubility and greater alkalinity of lime in comparison with limestone. This means that an effective removal of SO_2 from the flue gas is more readily performed with lime than with limestone.

On account of its chemical properties, the technically developed lime processes were earlier commercially available than those based on the use of limestone. On the other hand, the operational experience acquired with the lime based scrubbers could be advantageously applied for the conception and the development of the limestone based desulfurization processes.

Fig. 2 shows a simplified flow sheet of the flue gas desulfurization technoology as it was offered during the "seventies" in Japan, USA and partly in the Federal Republic of Germany.

This (old) technology is characterized by the following features:

- flue gas desulfurization and subsequent oxidation of calcium sulfite to gypsum are carried out in separate vessels; addition of sulfuric acid takes place to improve the oxidation conditions;

- separation of the solids is carried out by centrifugation in the cases that saleable gypsum has to be produced as the final product. Otherwise the gypsum slurry with a high water content is directly dumped;

- before passing to the chimney, the purified flue gas is reheated with additional energy (steam, natural gas, oil...).

The present state of the technology for the desulfurization of flue gases by limestone scrubbing is shown in Fig. 3.

The (new) flue gas desulfurization (abbreviated FGD) technology based on limestone scrubbing distinguishes itself from the older standards by the following improvements:

- the reaction chain aims at the production of gypsum for industrial use as the final product and the oxidation step is now integrated in the flue gas scrubber;

- simplification of the gypsum dehydrating system by application of continuous filters (vacuum belt filter, vacuum drum filter, screen bowl centrifuges) with hydrocyclones being placed upstream of the filters;

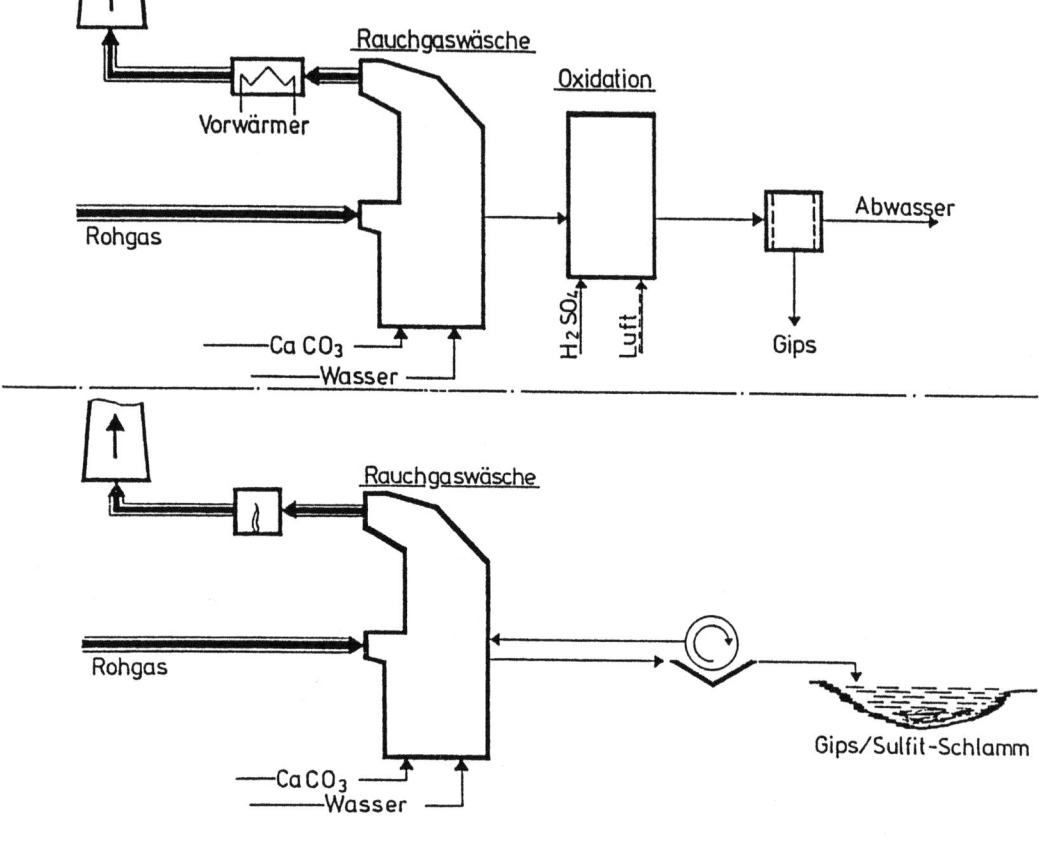

Fig. 2
Simplified FGD block diagram as offered from 1970 to 1980

reheating of the purified flue gases by Ljungstrom type gas-gas heat exchangers. However it has to be noted that this is not a specific feature of limestone processes alone.

2.2 Process description
A more detailed illustration of the features of the FGD technique by lime/limestone scrubbing is given by the simplified scheme in Fig. 4.

The flue gas coming from the boiler is first cooled down by the gas-gas heat exchanger and flows through the absorber spray tower. In most cases the flow in the reactor is countercurrent. Here, the following process steps are taking place simultaneously:

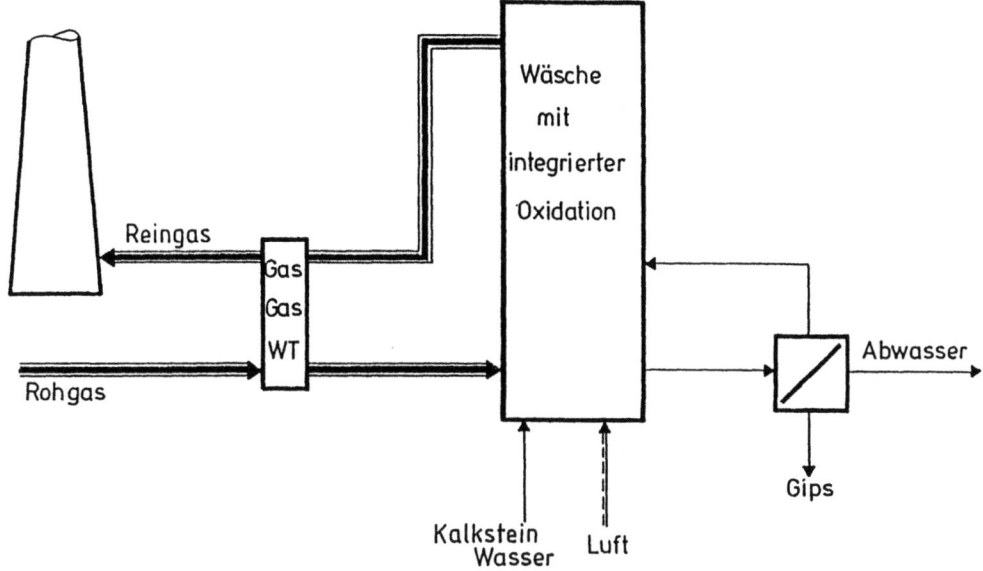

Fig. 3
Simplified FGD block diagram as offered since 1980

- absorption of the pollutant gas in the scrubber suspension;
- neutralization of the suspension by limestone;
- oxidation of intermediate neutralization products to gypsum;
- crystallization of the gypsum;
- separation of the flue gas from the scrubber liquid suspension.

Similar to the hydrated lime process, the flue gas is brought into contact with the scrubbing liquid in a spray tower. In this process, the acidic reacting SO_2 is absorbed. The scrubbing liquid can be neutralized with limestone where $CaSO_3$ (sulphite) and $Ca(HSO_3)_2$ (Bisulphite) occur as intermediate products. A small amount of oxygen is also absorbed from the flue gas in the scrubbing suspension, sufficient to convert the intermediates into $CaSO_4$. This oxidation reaction is not complete by itself, so that additional air is bubbled through the scrubbing liquid as an additional source of oxygen. The resulting reaction product ($CaSO_4$) is only partially soluble and precipitation of $CaSO_4.2H_2O$ (gypsum) occurs.

Separation of the gypsum from this suspension is normally carried out with the aid of a hydrocyclone/belt filter separation stage. This process makes it possible to

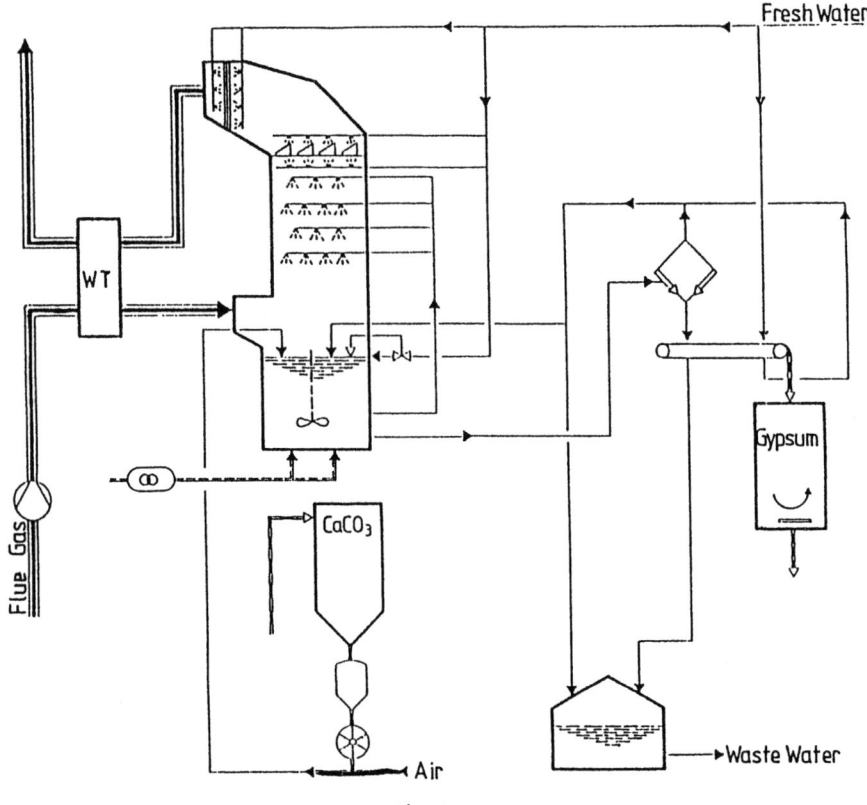

Fig. 4
Simplified flow sheet of a wet FGD

produce a particularly coarse-grained gypsum, as the fine particles are separated from the gypsum suspension and recycled back into the scrubber circuit.
Coarse-grained gypsum can be more efficiently dewatered than fine-grained gypsum, so that lower standards of moisture and chloride content can be maintained. This is particularly desirable when the gypsum has to be utilized commercially.

The total process chain is governed by three main control systems:

a) *Lime or limestone addition*
The quantity of the lime or limestone needed is nearly stoechiometric to the amount of SO_2 to be neutralized. In case of the use of limestone an excess of about 2 to 10% is necessary caused by the smaller solubility of this chemical in comparison with lime. The feed control occurs by determination of the incoming SO_2 flow rate by separate measurement of the SO_2 concentration and the flue gas flow rate, followed by calculation of the required limestone

feed rate. This calculated value will be corrected on the basis of the measured SO_2-content of the purified flue gas or alternatively by the pH-value of the absorber liquid, which is normally kept between 5.0 and 5.6.

b) *Liquid level in the absorber circuit*
By control of the liquid level in the absorber bottom, the water balance of the process is controlled. The main quantities of fresh water to be added to the process are:
- washing water for the droplet demisters in the scrubber;
- washing water for the gypsum in the filtration system.

The main quantities of water leaving the process are:
- the evaporation of water in the scrubber; by this the flue gas is cooled down to its saturation point;
- the waste water stream: this effluent is necessary to remove the soluble chloride salts from the scrubbing suspension. Chloride enters the process as HCl contained in the flue gas.

c) *Solid concentration of the absorber liquid*
Measurement of the density of the scrubbing liquid controls the discharge of the crystallized gypsum. A small effluent stream is led to the gypsum separation sytem. This bleed stream is very small in comparison with the liquid circulation rate inside the scrubber. The main circulation rate depends mainly from the SO_2-inlet concentration of the flue gas and the desired SO_2-removal efficiency. The so-called "liquid to gas ratio" normally varies from 5 to 25 liter per m^3.

Fig. 5 gives an example of the main process data for a typical FGD plant for a hard coal fired power plant boiler. The required electric energy is mainly needed by the fan for the compensation of the pressure drop of the flue gas flow through the plant (i.e. heat exchanger, scrubber, flue gas channels...) and the slurry circulation pumps. The quantity of waste water and consequently the need of fresh make-up water can be reduced by increasing the chloride content of the scrubbing liquor. Normal values for the concentration of chlorides in the scrubber liquid (and thus in the waste water) are 10 to 20 g/l.

3. Lime and gypsum quality

3.1 Limestone
Limestone is a natural mineral which means that its physical and chemical properties vary with the area where it is mined. Normally, a FGD plant will be designed in such a way that the natural regionally occurring limestone can be applied as the neutralization agent.
It is advantageous to use a limestone quality with a high reactivity. In general it can be said that the reactivity of the types of limestone increases in the following order:
    Trias - Devon - Jura - Chalk.
The impurities contained in the limestone can sometimes have a positive effect on the reactivity.

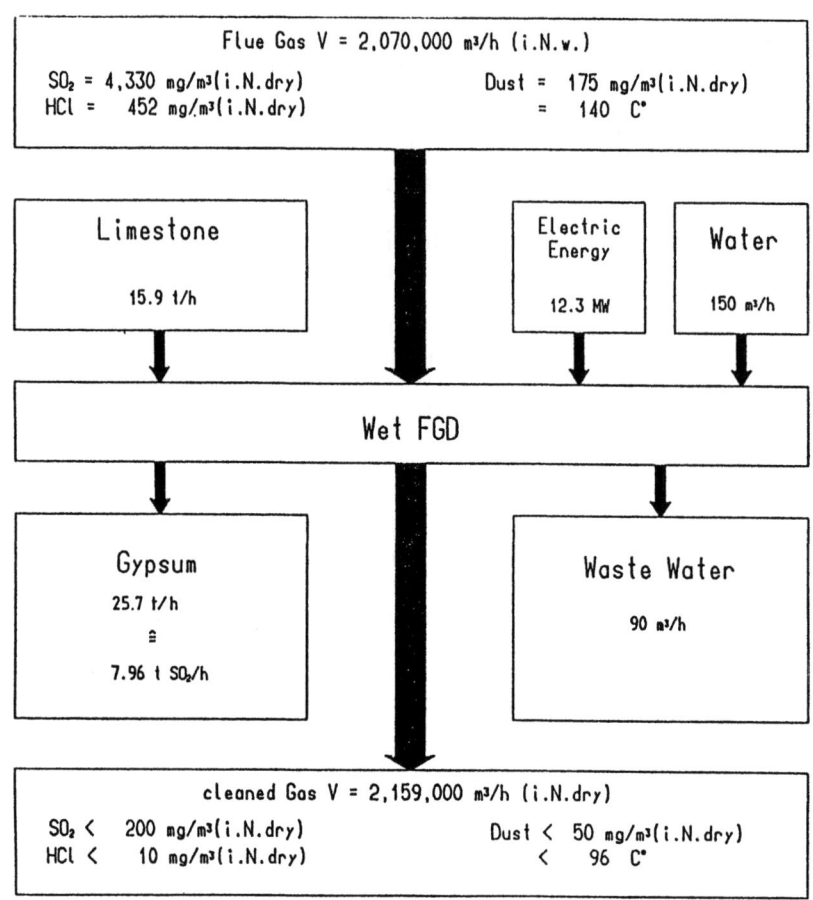

Fig. 5
Mass and energy balances of a wet FGD system
(approximately 600 MWe)

Fig. 6 give a typical example of the laboratory tests for limestone reactivity.

However, from this general trend no conclusion as to the applicability of the limestone can be drawn. The apparent reactivity is positively influenced by higher grinding fineness and reduced by the increase of the $CaCl_2$-content of the scrubbing liquid. On the whole, the question of the limestone quality to be used depends on the overall design of the flue gas desulfurization system. As a general rule, processes are designed for a limestone grinding fineness of:

$$90\% \leq 90\ \mu m, d' \leq 22\ m$$

and a purity ($CaCO_3$-content) of more than 95%. The parameter d' corresponds to a sieve residue of 36.8%. The purity is specified for a gypsum quality which should contain more than 90% $CaSO_4.2\ H_2O$. To reach this purity, a particle content of the incoming flue gases of less than 80 mg/Nm^3 is necessary.

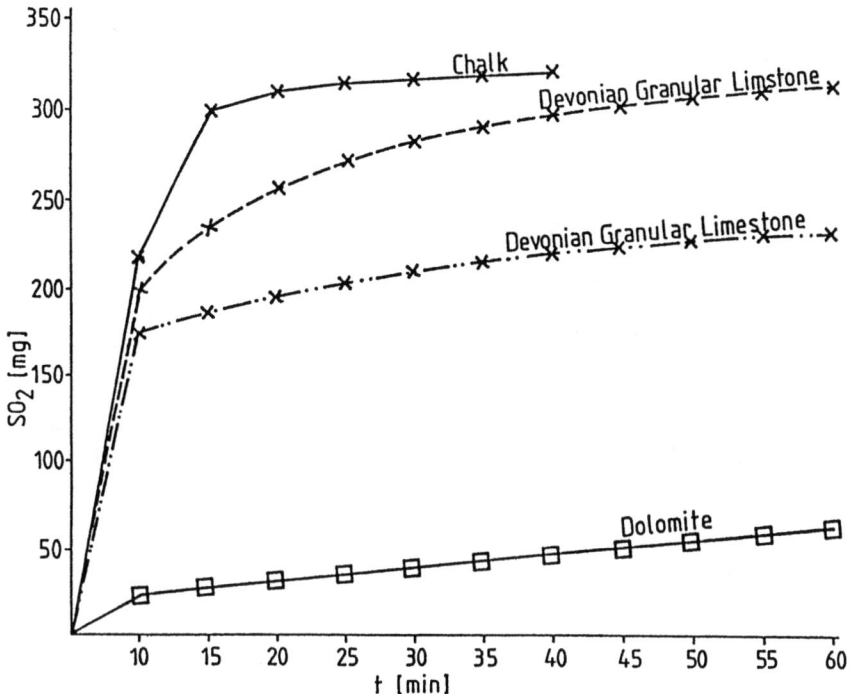

Fig. 6
Reactivity profiles of different limestone types

.2 Gypsum
.2.1 *Quality* The specifications of the gypsum industries for the purity of gypsum from a FGD is the following:

Humidity	$\leq 10\ \%$
$CaSO_4 \cdot 2\ H_2O$	$\geq 95\ \%$ *
MgO soluble	$\leq 0.1\ \%$
Chloride soluble	$\leq 0.01\%$
Na_2O soluble	$\leq 0.06\%$
Sulfur dioxide (SO_2)	$\leq 0.25\%$
pH value	5 - 9
Colour	white (grade $\geq 80\%$)
Odour	neutral
Toxic components	none

* = Content of inert components may be higher than 5% because - similar to natural gypsum - these components do not mainly effect the quality.

Normally it is no big problem to meet these specifications, but a certain awareness is necessary to the specifications for the colour (white $\geq 80\%$) and the humidity (less than 10%).

The colour is mainly influenced by
- particle content of the flue gas at the absorber inlet and by the carbon content of these particles;
- impurities of the limestone.

The countermeasures consist either of the installation of a prescrubber and/or the use of lime instead of limestone in case that the specifications for a sufficient white colour cannot be met. Otherwise there is the alternative to pass the second grade gypsum to the cement industry, where the colour is less important.

The humidity is controlled by the dewatering process. The size and the shape of the gypsum crystals have an important influence on the humidity: big and rectangular crystals favour the dewatering process. The crystal properties can be controlled by
- retention time of the solids in the scrubber;
- solid concentration in the scrubber liquid;
- concentration of $CaCl_2$ or Na_2SO_4 in the scrubber liquid.

Fig. 7 shows typical results of the crystallisation of gypsum in a laboratory test by addition of $CaCl_2$ and Na_2SO_4. It is clearly demonstrated that $CaCl_2$ has an adverse effect on the quality, whereas the presence of Na_2SO_4 has a positive influence.

3.2.2 Handling of gypsum Gypsum obtained from flue gas desulphurization (FGD gypsum) normally has such a high degree of purity that a processing plant for "chemical purification" is not required. Such additional purification steps are usually necessary for waste gypsum from the phosphoric acid industry

The FGD gypsum can be delivered to the consumers, such as gypsum- or cement works, as it is obtained from the filter units or after drying and pelletizing.
The specific end user finally decides about the specifications for the special quality of gypsum needed for his purposes. A gypsum factory in the vicinity of a power station may also have the facilities to accept the wet gypsum power as such, although in such cases loading and transport might pose problems.
Nowadays, a substantial part of the FGD gypsum produced in Germany is used for the production of wallboard gypsum.

A wider variety of uses can be achieved by offering pelletized gypsum instead of powder. The silos available in the gypsum and cement industries are usually suited for products with a size range of 0 - 25 mm and could therefore also be used for the storage of pelletized gypsum. Moreover, pelletized gypsum can be transported in open or covered lorries in contradistinction to the transport of gypsum powder which requires the application of fully enclosed special-type bulk transporters.

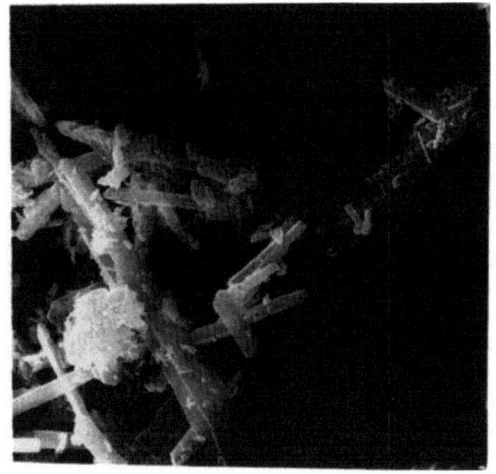

Test A: addition of $CaCl_2$ Test B: addition of Na_2SO_4

Fig. 7
Crystallization of gypsum with different additives
Enlargement: 2000

The initial strength of the dihydate pellets required for silo storage can be achieved by pelletizing in a roller press similar to those used for the production of ovoid coal briquettes. The roller press technique requires pre-drying of the gypsum before pelletizing. The presence of surface moisture of the FGD gypsum will have an adverse effect on the pressing process whereas also the strength of the produced pellets will decrease with increasing moisture content.

The drying process before pelletizing has to produce a product with absolutely uniform free moisture content, next to zero, whereas on the other hand not too much water may be removed so that part of the calcium sulfate dihydrate is transferred into hemihydrate or anhydrite. This is necessary with a view on the storage of the powder before the pelletizing process as well as to achieve the required silo strength of the pellets.

Such stringent drying conditions can be met by the use of special equipment, for example by a continuous contact dryer. The FGD gypsum passed through this type of dryer and pressed to pellets of any shape, beit briquette, cushion, cigar or other, can simply be transferred into a closed silo system for storage. From there it can be discharged without problems if and when required. Fig. 8 shows the principles of a pelletizing process.
In 1990 the production of FGD gypsum in West Germany will amount to 3.3 million tons per year. At the other side there is a consumption of gypsum in the

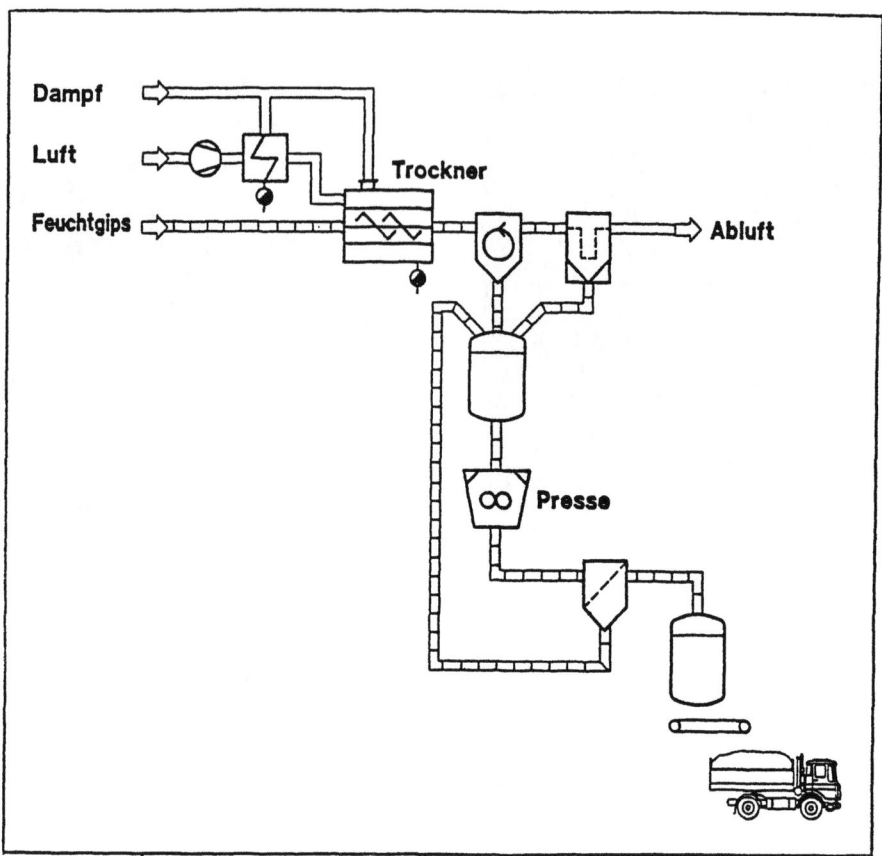

Fig. 8
FGD Power plant Heilbronn Block 7
Gypsum drying and pelletizing

construction material industry of about 5 million tons per year. Except for the gypsum from lignite fired power plants (about 1 million ton per year), which has to be dumped in the open air lignite coal mines, all produced quantities are used:

- in the cement industry (gypsum content of cement up to 3% for regulation of the hardening of concrete);

- for wallboard production in the gypsum industry;

- for stiffening tunnel walls in the mining industry

- as raw material for other gypsum applications.

4. Costs

The installation of FGD in nearly all power plants in West Germany (new plants and retrofits) was carried out in a period of 5 years after the legislation was settled in 1983. This means that a capacity for 38 000 MW_{el} had to be installed. The total costs for this effort amounted to 14.2 billion DM, which means that the average installation cost is 370 DM/kW_{el}. This is an astonishingly high figure, keeping into account that 90 % of this capacity consists of lime/limestone plants.

However, it is quite important to recognize that there is a large difference between the costs for a retrofit and the construction of a new plant. It can be estimated that retrofits causes about 30% higher investment costs, needed for additional civil engineering, modifications of the power plant, removal of existing flue gas ducts and installation of other additional flue gas ducts.
The average investment costs for new plants are in the order of 260 DM/kW_{el}. This value depends naturally of the size of the plant and of the amount of additional equipment added to the basic system. Such additions may exist of special waste water treatment units, special gypsum handling and/or limestone preparation, the installation of pre-scrubbers etc.

Fig.9 gives a detailed survey of the costs for a new plant, calculated by the power plant operator. In this particular plant (Heilbronn Block 7) a large number of additional equipment like limestone mills, gypsum pelletizing and wate water treatment was installed.

The operation costs of the plant (including depreciation of capital costs, labour costs, maintenance etc.) are given in Fig. 10. The costs per kW_{el} depends heavily on the operation time of the power plant per year. The reason for this fact is that the direct operation costs like limestone consumption, water and electricity are relatively low in comparison with the fixed operation costs.

The credit obtained from the selling of the by-product gypsum is negligible. The direct income from the sale is practically counterbalanced by additional costs for producing a better gypsum quality (pelletizing), for additional storage costs or for other, similar affairs.
At the other hand, if the costs of the produced gypsum were to be considered as a function of the total FGD costs, the price of the produced gypsum would be as high as 1000 DM/t. Natural gypsum costs approximately 10 DM/t. However, this comparison is a nonsense, the real sense of a FGD is the improvement of the environment and it is very difficult to assign a definite value to this asset.

5. Examples of process modifications

The usual design of FGD installed in hard coal or lignite fired power plants is based on SO_2-removal efficiencies between 90 and 95% and/or SO_2-concentrations at the scrubber outlet between 200 and 400 mg/Nm^3.
The efficiency of a one-stage scrubber on the basis of lime or limestone is limited because of two reasons:

Planung	2%
Sonstiges	2,5%
Kalksteinmahlanlage	2,5%
Abwasseraufbereitung (komplett)	4%
Bauteil	8%
Gipsaufbereitung	8%
Stahlbau	11%
E- und L-Technik	11,5%
Wiederaufheizung	12,5%
Verfahrenstechnik	38%
Investitionskosten der REA zum IBN-Zeitpunkt	100% 270 DM/kW

Fig. 9
FGD Power plant Heilbronn Block 7
Investment costs

- limestone is a weak base, and therefore the reactivity of the scrubber liquid is also low;

- in connection with the above, the scrubber is sensitive to a non-uniform liquid-gas distribution over the tower cross-section in case of very low SO_2-concentrations in the outlet.

It is, in principle, possible to increase this efficiency by the installation of two scrubbers in series. A cheaper and more effective solution is the use of a strong base like caustic soda (NaOH) in the second stage, see Fig 11.

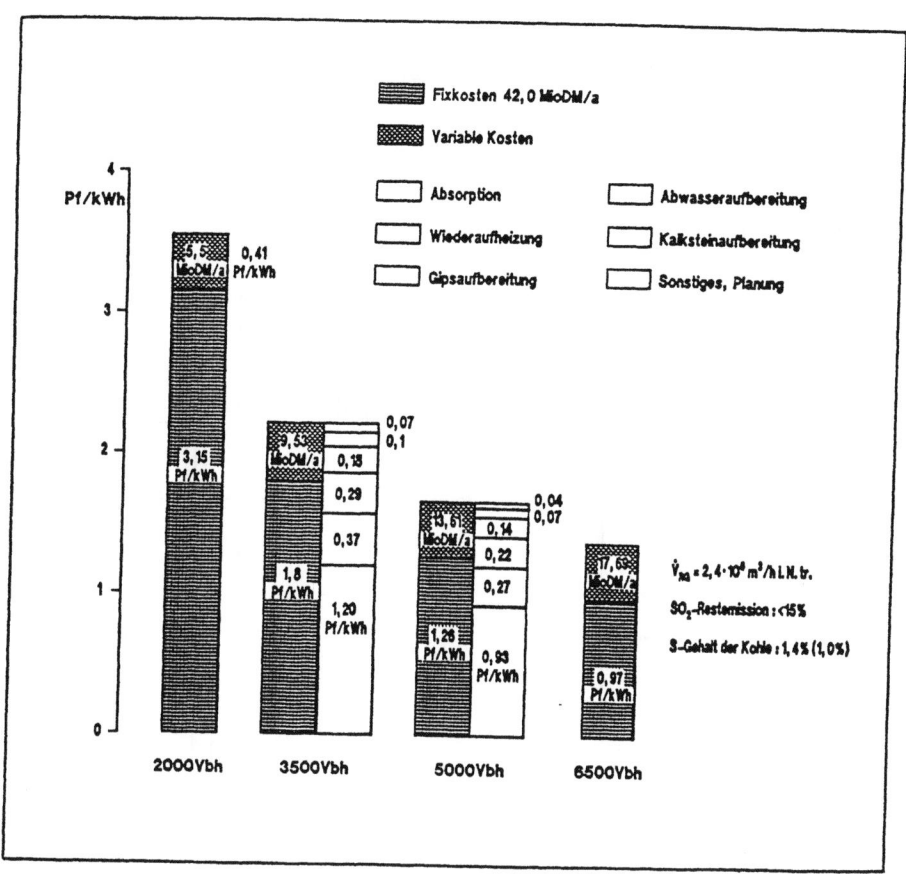

Fig. 10
FGD Power plant Heilbronn Block 7
Operation costs

The additional scrubbing stage can be constructed as a venturi type scrubber, because a venturi is relatively cheap and requires less space.
The needed quantity of additive, i.e. the expensive caustic soda, is low, because only the SO_2-outlet concentrations of the first scrubber (the limestone scrubber) have to be neutralized by the soda.
The neutralization with soda produces Na_2SO_4, which has to be removed from the process. The product is recycled into the bottom of the first scrubber. Then, the following reactions take place:

1) In the main scrubber (first scrubber):

$$SO_2 + \tfrac{1}{2}O_2 + H_2O + CaCO_{3\,s} \longrightarrow CaSO_4 \cdot 2H_2O_{\,s} + CO_{2\,g}$$

$$2\,HCl + CaCO_{3\,s} \longrightarrow CaCl_{2\,sl} + H_2O + CO_{2\,g}$$

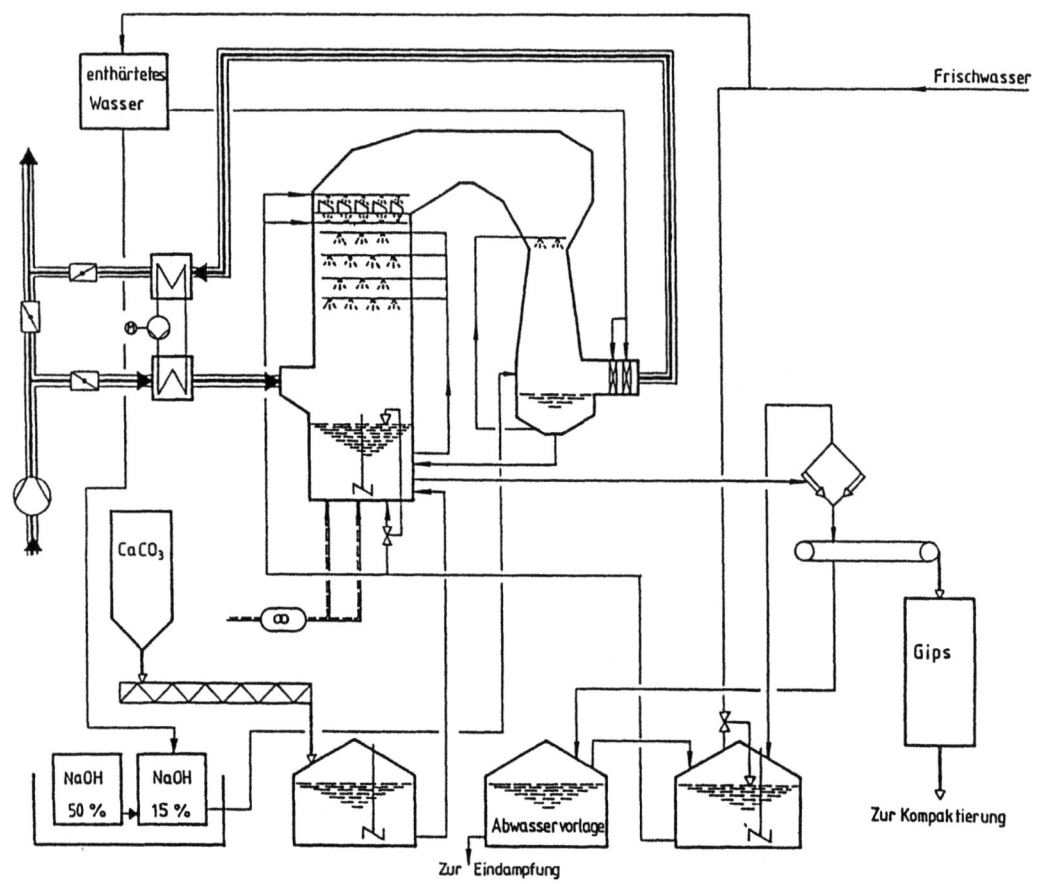

Fig. 11
Combined $CaCO_3$ - NaOH FGD

2) *In the second scrubber:*

$$2 NaOH_l + SO_2 + \tfrac{1}{2} O_2 \rightarrow Na_2SO_{4\,sl} + H_2O$$

3) *In the bottom of main scrubber*

$$CaCl_{2\,sl} + Na_2SO_4 + 2 H_2O \rightarrow CaSO_4.2H_2O_s + NaCl_{sl}$$

The soluble $CaCl_2$, formed by the removal of HCl from the flue gas, reacts with the soluble Na_2SO_4 to form the insoluble gypsum. In the scrubber solution remains NaCl, which can be tolerated at high concentrations because it does not exert negative influences on the SO_2-removal efficiency or gypsum crystall size, this in contradiction with $CaCl_2$, as indicated before. This means that the rate of waste water production remains low and the only external effect of the proposed combination is that the waste water now contains NaCl instead of $CaCl_2$.

Other process modifications - for example the use of calcined dolomite instead of pure limestone are also possible. Dolomite ($CaCO_3$ + $MgCO_3$) is a natural product. It is possible to produce valuable $Mg(OH)_2$ at a high degree of purity by the so-called "combined lime-magnesium process", developed by the Austrian company Refractories Consulting Engineering in cooperation with Deutsche Babcock Anlagen AG.
These examples of variations of the lime-gypsum process may demonstrate the advantages of the simplicity of this process, which makes it possible that adaptations to specific technical or economic conditions are easily carried out.

6. References

[1]   B. Schönbucher, Th. Poensgen and H. Fahlenkamp, " Aufbau und Funktion der Rauchgasentschewelungsanlage im Kraftwerk Heilbronn der EVS AG", *VGB Sondertagung "Neue Steinkohlen-Blöcke am Neckar"*, 6-7 November 1986 and VGB Kraftswerktechnik 3/1987

[2]   H. Fahlenkamp, "Recent developmnets in West-Germany's limestone based FGD Technology", *ASTM/IEEE Power Generation Conference*, Paper No. 86-JPGC-EC-18, Portland, 19-23 October 1986

[3]   Paper of the Deutsche Babcock Anlagen AG, Krefeld: "Desulphurization of flue gas"

[4]   Energiewirtschaft und Technik Verlaggesellschaft mbH .,"Entschwefelung (overview)", July 1988

[5]   B. Schärer and N. Haug, "Bilanz der Grossfeuerungsanlagen-Verordnung", Staub-Reinhaltung der Luft, 50 (1990) 139-144

THE WELLMAN LORD PROCESS

U. Neumann
Davy McKee AG
D-6000 Frankfurt/Main
Federal Republic of Germany

1. Introduction

The Wellman Lord process is a regenerative process. The basic principles of the process consist of removing highly diluted SO_2 from the flue gas in the absorption section and then turning it into rich SO_2 gas in the regeneration section. The recovered gas is directly processed into the end products which are alternatively elemental sulphur produced by the Claus process, or concentrated sulphuric acid, or liquid SO_2. These end products meet the highest quality standards.

The process does not generate waste-disposal problems and does not require any raw materials that reappear in the final product. The pollutant SO_2 is converted into valuable and saleable products with guaranteed long-term utilization. The process reduces the remaining environmental pollution to a minimum, protects natural resources, and avoids logistic problems.

The process is highly flexible and can be used in connection with flue gases generated by the combustion of all relevant fossil fuels. It is particularly favoured when flue gases with high or fluctuating SO_2 content are to be processed. In this respect, the technology is highly suited for the desulphurization of flue gases emitted by lignite-fired power stations.

D. van Velzen (ed.),
Sulphur Dioxide and Nitrogen Oxides in Industrial Waste Gases: Emission, Legislation and Abatement, 111–137.

2. Process Fundamentals

The process employs a sodium sulphite solution in the absorption-regeneration circuit. The solution selectively removes SO_2 from the flue gas, and sodium bisulphite is formed. During the absorption and subsequent desorption processes, the pH of the solution varies only in a small pH range as it is typical for a buffering system.

The absorption reaction is reversed in the regeneration section. By thermally regenerating the product solution in an evaporator crystallizer, the absorbed SO_2 is released again and, after the water vapours are removed by cooling, it emerges as SO_2 rich gas. The resulting condensate is used to redissolve sodium sulphite crystallized during regeneration. The fresh solution gained is ready again for SO_2 absorption, and the circuit is closed.

For a better understanding of the process, it will be helpful to know more about the solubility properties of the main components in the Wellman Lord solution: The solubility of sodium sulphite in water is quite high. But it is considerably exceeded by the solubility of sodium hydrogen sulphite ($NaHSO_3$), as it is shown in **FIGURE 2**.

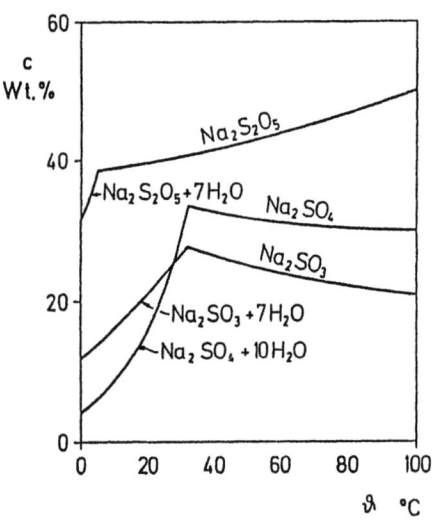

FIGURE 1

Solubilities of
Na_2SO_3, $Na_2S_2O_5$
and Na_2SO_4 in Water

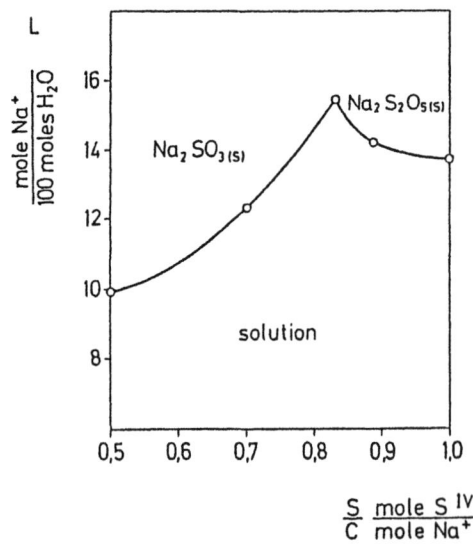

FIGURE 2

Equivalent Solubilities L
in the System Na_2SO_3-
$Na_2S_2O_5$-H_2O
at 45 °C

An important feature of the process is the high capacity of the circulating solution for absorbing SO_2 which is based upon the favourable sulphite solubility and is furthermore achieved by operating the Wellman Lord system with solutions close to saturation.

When SO_2 is removed from flue gases by the Wellman Lord process, the overall solubility of the Wellman Lord solution, in terms of sodium dissolved, increases progressively with the amount of SO_2 absorbed (see **FIGURE 2**). As a result, no solids are formed in the Wellman Lord absorber or can be precipitated out of the solution.

Absorption of SO_2 takes place in a three-stage or four-stage absorber. The absorbing solution passes through the absorber in a countercurrent flow to the flue gas and is successively loaded with SO_2.

Thermal regeneration is effected by evaporating the absorber product solution in an evaporator crystallizer. At steady state operation of the evaporator, sodium hydrogen sulphite, or its dewatered form, $Na_2S_2O_5$, is highly enriched in the liquid phase, with the result that the solubility of sodium sulphite is reduced accordingly.

The high sodium hydrogen sulphite concentration in the evaporator liquid phase favours its decomposition. While water is progressively evaporating, SO_2 is released, and sodium sulphite crystals are continuously being formed. The vaporized water serves as a carrier for discharging SO_2 from the system. The fresh absorber feed solution is regained by dissolving the crystallized sulphite in the condensate which results from cooling down the vapours, and is ready for being used for further SO_2 absorption. The various process steps are illustrated in **FIGURE 3**.

The SO_2 gas recovered in the regeneration section is very pure and highly concentrated. It is directly processed either to sulphur in a combined thermal reduction and catalytic Claus plant, or to sulphuric acid, or to liquid SO_2. The SO_2-containing off-gases from the Claus or sulphuric acid plants are returned to the FGD absorber.

A small amount of sulphite is oxidized in the absorber by oxygen contained in the flue gas. The sodium sulphate formed is separated from a side stream of the circulating solution and further processed to a pure sodium sulphate product. A small evaporator crystallizer is used for sulphate separation. It also serves for preconcentrating the evaporator feed solution.

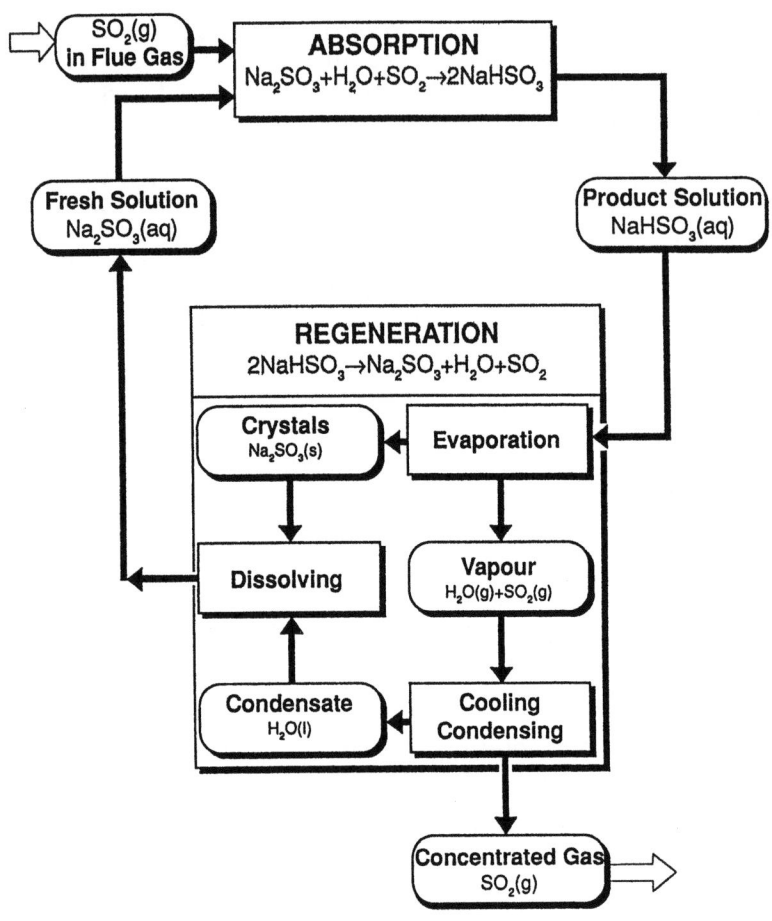

FIGURE 3

Absorption and Regeneration Circuit

As the Wellman Lord process results in high quality products, other harmful components in the flue gas like HCl, HF, and fly ash fines are effectively removed in a prescrubber and are consequently accumulated in the waste water. The purge water flow is minimized in the process.

The level of impurities in the circulating solution is controlled by purging a small amount of evaporator mother liquor. The impurities originate from traces of the above flue gas contaminants not removed in the prescrubber and from minor side reactions in the process. The losses in sodium compounds are balanced by adding sodium hydroxide solution as alkaline make-up to the circuit.

FIGURE 4 shows a simplified diagram of the Wellman Lord system as described in the above paragraph.

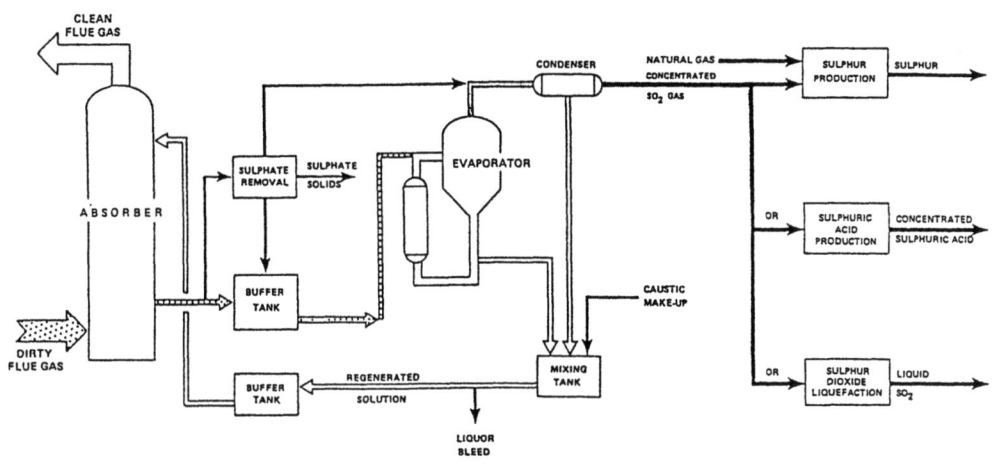

FIGURE 4

Simplified Flow Sheet of the Wellman Lord System

3. Process History and References

The process was developed by the Wellman Lord Company in the late 60's. The company was located in Florida and had their main business with the Florida phosphate fertilizer industry which had a growing demand for sulphur. The development was initiated at a time when the sulphur prices were high and the energy prices still very low. The objective was to open a new source of sulphur for the economical production of sulphuric acid for use in the fertilizer industry.

This objective of the process development could never be met due to drastic changes in the economic conditions. However, the process was available at the right time when SO_2 emissions were limited by environmental regulations in the US and in Japan.

The first commercial plant for recovering SO_2 from sulphuric acid tail gases went onstream in the U.S. in 1970, followed by two Wellman Lord plants in Japan, added to an oil-fired boiler and a Claus plant, in 1971. Within a period of twelve years, altogether 30 Wellman Lord plants were built in the U.S. and in Japan for desulphurizing flue gases with an accumulated flow rate of 17×10^6 Nm^3/h.

The Wellman Lord activities were then transferred to Europe. The first Wellman Lord plant ever built in Europe was installed at the Schwechat refinery of ÖMV in Austria. It was needed for desulphurizing flue gases from heavy oil burning and refinery off-gases. Five more plants were then built in Germany in the 80's, all for coal or lignite-fired power stations. The European plants added a flue gas flow of in total 7.2×10^6 Nm^3/h to the Wellman Lord plant inventory. Information on these plants has been compiled in **TABLE 1**.

TABLE 1

Wellman Lord Plants in Europe

Client	Location	Fuel	Product	Flue Gas Flow Rate Nm^3/h	SO_2 Content g/Nm^3 (dry)
ÖMV	Schwechat/ Austria	heavy residues	sulphur	2 x 300,000	4.6 - 6.3
BASF	Ludwigshafen/ Germany	hard coal and sludge	liquid SO_2	2 x 570,000	1.6 - 2.8
BKG	Marl/Germany	low-grade coal	liquid SO_2	2 x 600,000	4.5
BKB	Helmstedt/ Germany				
	- Buschhaus	high salt lignite	sulphur	2 x 807,000	10 - 19
	- Offleben C	lignite	sulphur	2 x 800,000	6 - 12
KAB	Rummelsburg East Berlin	lignite	liquid SO_2	1,048,000	4.0 - 6.2

The Wellman Lord system owned by Davy is the only regenerative technology which has continuously built up experiences over the last two decades with a wide variety of waste gas sources and a wide range of gas flows. Operational experiences are available from 37 plants representing an accumulated operating time of about 375 years. Sulphuric acid was the final FGD product of most of these plants (20), followed by sulphur (13 plants) and liquid SO_2 (4 plants). The total SO_2 recovery capacity of these plants exceeds 850,000 tons/year. **TABLE 2** surveys the plant experiences with different waste gas sources.

TABLE 2

Plant Experiences with Different Waste Gas Sources

Number of Plants	Waste Gas Source	Flue Gas Flow Nm^3/h
16	Oil	7,700,000
5	Coal	10,760,000
3	Lignite	4,260,000
1	Petroleum Coke	920,000
12	Chemical Plants	690,000
37		24,330,000

4. Overall Plant Design

A Wellman Lord plant for flue gas desulphurization comprises three main areas:

- The absorption or flue gas treatment unit which includes reheating, prescrubbing and SO_2 removal.

- The regeneration unit including sulphate separation from the circulating solution.

- The downstream area for processing the intermediate products of the Wellman Lord plant to final products and by-products.

In addition, a treatment unit for the prescrubber purge water is required.

The flue gas treatment unit is directly adjunct to the power station and operates according to the boiler load schedule. Due to the **Clean Absorber** concept, the absorption unit operates very reliably and with a minimum down time.

Intermediate storage tanks for both absorber feed and product solutions serve to isolate the absorption area from any disturbances in the regeneration system and vice versa. Variations in flue gas flow and SO_2 content thus have little effect on the regeneration system, which in turn may be shut down for repair or scheduled maintenance without interrupting SO_2 removal. As both plant areas are only linked by two pipes for the circulating solutions, it can be easily arranged to install the regeneration plant at some distance from the flue gas treatment unit.

The option to locally separate the regeneration plant from the power station is an important feature, especially when the FGD plant has to be retrofitted to an existing power station. At a number of Wellman Lord reference plants, the regeneration facilities were installed at distances of some hundred meters or more from the flue gas treatment unit. A pipeline of 7 km in length, for instance, connects the absorption unit serving the Offleben Block C power station with the central Wellman Lord regeneration facilities located at the Buschhaus power station of BKB, West Germany.

By intermediate storage of the prescrubber purge water it is achieved to operate the waste water treatment unit and the flue gas treatment section independently of each other. The waste water unit is designed to meet the relevant environmental standards.

5. Process Units

As it is essential for a commercial process of long-term application and high market potential, the Wellman Lord process has been continuously improved over the 20 years period since the first commercial Wellman Lord plant went on-stream. The following description of the process units reflects the actual development and design status of the Wellman Lord system.

5.1 Flue Gas Treatment Section

Reheating is performed either by rotating heat exchangers of the Ljungström type or by stationary heat exchangers linked by a heat transfer medium. Due to the leakages of the rotating system, the liquid-linked system is preferred at high SO_2 levels in the raw flue gas. Davy have operational experiences with various systems and materials.

Prewashing of the flue gas takes place in a one or two packed stages **prescrubber**. The new design prescrubber is installed in both BASF plants at Ludwigshafen and Marl and operates reliably. Two stages are required for effectively removing HCl from coal based flue gases. The prescrubber design provides for reduction of the waste water purge from the prescrubber to such an extent that even the zero purge waste water concept can be realized at reasonable cost.

The **absorber** design has been changed from valve trays to packings. The new design results in a considerably reduced pressure drop even at better absorption efficiencies and allows the absorber to effectively operate also at low boiler load conditions.

The **flue gas blower** is installed on the clean gas side. This position has been successfully proven on a long-term basis in all European plants. As the Wellman Lord solutions only contain soluble salts, there is no risk of any serious and long-term scaling at the blower blades.

5.2 Regeneration System

For the regeneration of Wellman Lord solutions, single or double effect evaporator crystallizers were normally used. Evaporation is performed under vacuum at reduced temperatures for the reason of limiting the formation of side products, especially thiosulphate.

One of the major improvements in the regeneration area relates to energy savings. The latest Wellman Lord reference plants were equipped with mechanical vapour recompression. By utilizing the compressed evaporator vapours as heat source for the evaporator operation, considerable steam savings have been achieved with the result that no steam import was required any more for normal operation of the whole flue gas desulphurization plant.

The following modifications also contributed to the overall energy savings achieved:

-   high temperature sulphate separation (see para 4.3),
-   vapour condensation and condensate stripping,
-   crystals separation by centrifuges.

The design of the regeneration unit with mechanical vapour recompression and high temperature sulphate separation is outlined in the basic flow diagram, **FIGURE 5**.

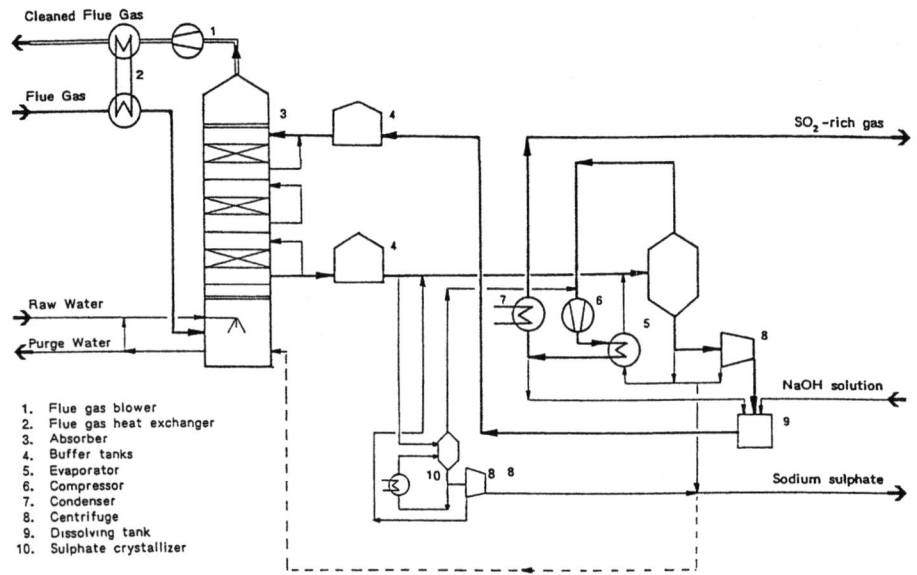

FIGURE 5

Wellman Lord System with
Mechanical Vapour Recompression

It is evident from the diagram that crystallization by evaporation is the key unit operation in the regeneration section. There is a high number of such crystallizers being operated all over the world with plant sizes covering a wide range of capacities. These plants are mainly related to the saline and soda industries as well as to sodium sulphate recovery, and many of them, especially the larger ones, are equipped with mechanical vapour recompression. Consequently, the Wellman Lord system applies simple unit operations and highly proven and reliable equipment.

5.3 Raw Sulphate Separation and Processing

Sodium sulphate formed in the absorber by sulphite oxidation (see section 2) is removed from the Wellman Lord solution in a high temperature sulphate separation unit which is an integrated part of the regeneration plant.

The sulphate separation system consists of an evaporator crystallizer unit. It operates under special conditions favourable for sulphite separation, and simultaneously preconcentrates part of the feed solution to the regeneration section.

The mixed sulphate and sulphite crystals separated by a centrifuge can be optionally processed to a pure sodium sulphate product. This operation is required and even economically justified at high FGD plant capacities. The dissolved raw sulphate is first treated with air in order to oxidize the sulphite contained in the solution. The sulphate solution is then evaporated in an evaporator crystallizer. By separating and drying the resulting crystals, a sodium sulphate product is obtained which meets high quality standards. A pure sodium sulphate plant is successfully in operation at the Buschhaus FGD plant.

5.4 SO_2 Gas Processing Unit

The concentrated SO_2 gas recovered in the regeneration section can be converted to either elemental sulphur, or sulphuric acid, or liquid SO_2. The relevant technologies are in Davy's portfolio, and they are based on proprietary know-how and an extensive track record.

Sulphur is produced by the Redotherm-Claus Process (see FIGURE 6). The process requires:

- a thermal SO_2 reduction unit,
- a two-stage catalytic Claus unit,
- a tail gas incinerator.

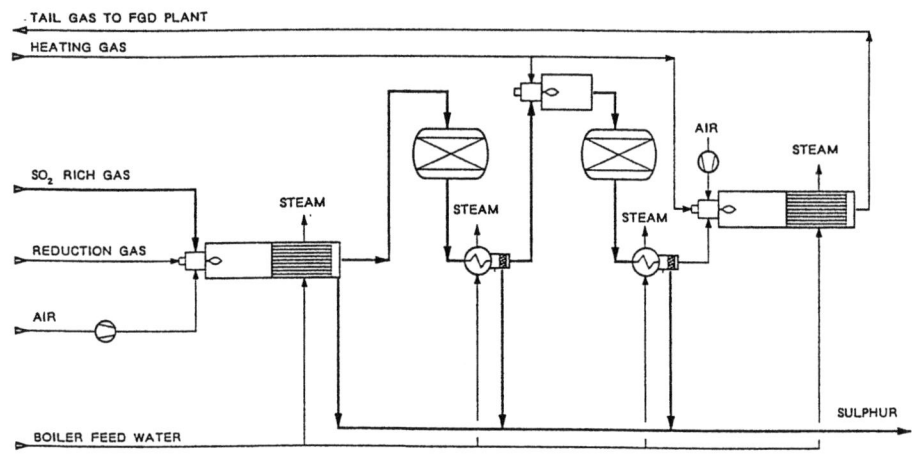

FIGURE 6

Sulphur from SO_2 Rich Gas by the Redotherm Process

Hydrogen or any reducing gas like town gas as well as natural gas can be used. If natural gas is used for SO_2 reduction, the modified Allied-Claus Process can be applied as an alternative. The process is based upon catalytic SO_2 reduction. As the tail gas is normally recirculated to the WL absorber, a tail gas abatement unit is not required.

Sulphuric acid is produced by the catalytic conversion process (see **FIGURE 7**). The unit consists of a drying tower for drying the SO_2 feed gas and the air required for oxidation, the SO_2 absorber and the catalytic converter. As the tail gas is also returned to the Wellman Lord absorber, there is no need for double absorption, and only the single absorption system is applied.

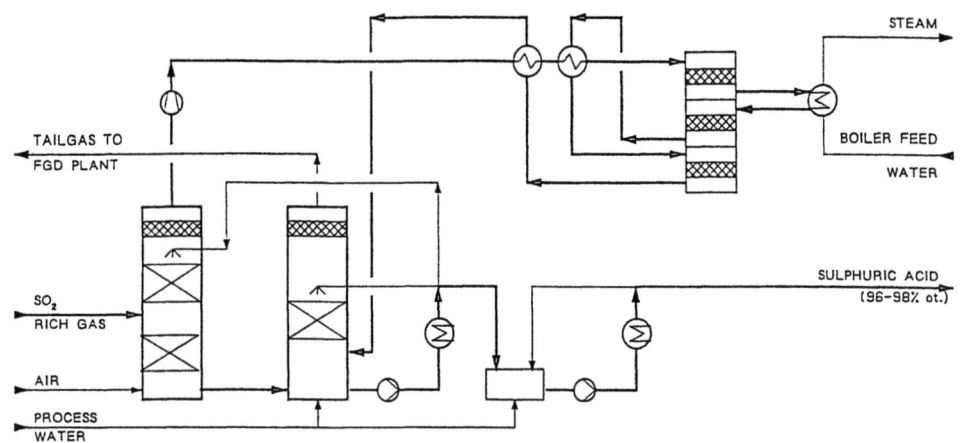

FIGURE 7

Sulphuric Acid from SO$_2$ Rich Gas

Processing the concentrated SO$_2$ gas to liquid SO$_2$ is another product option. The intermediate SO$_2$ gas is first dried by passing it through an adsorption unit, and then liquefied either at lower pressure by using a chilled medium for cooling, or alternatively at medium pressure at which cooling water can be employed for liquefaction. The liquid SO$_2$ produced is of outstanding quality.

5.5 Waste Water Treatment Unit

The acidic waste water is neutralized by lime or limestone which takes place in a two-stage or three-stage neutralization unit. Precipitation of calcium fluoride and gypsum is improved by recycling of solids to the preneutralization stage.

By accurately controlling the pH in the neutralization section as well as by adding specific agents, the heavy metals content is reduced to the extent required to meet the environmental standards.

The solids are accumulated in a sedimentation unit, and the resulting slurry is filtered for further dewatering of the solids. The residue has to be separately disposed. The treated waste water is discharged into public water systems.

As the prescrubber purge flow from Wellman Lord plants is kept to a minimum, the capacity of the treatment unit is accordingly small.

6. Products

6.1 Main FGD Products

As already outlined in section 2, sulphur dioxide removed from the flue gases is recovered in the Wellman Lord regeneration section as enriched and pure SO_2 gas. The high purity of the concentrated gas can easily be explained by the process route and the operating conditions applied.

The SO_2 absorption is very selective, and volatile contaminants do not really arrive in the circulating solution. Consequently the SO_2 and water vapours released from the regeneration unit are already quite pure, and any impurities still transferred to the vapours are then absorbed or dissolved in the condensate when the vapours have been cooled down.

In fact, the liquid SO_2 produced in commercial Wellman Lord plants just by liquefying the recovered and dried SO_2 gas, without any further purification steps involved, is of outstanding quality. In case of sulphuric acid and sulphur production, excellent product qualities are also achieved.

From the three possible products to be made from the intermediate SO_2 gas, sulphur has the most favourable storage and handling properties, but due to the expenses for the reducing gas required, the overall production costs are the highest. On the other hand, liquid SO_2 has the lowest production cost, but due to the noxious character of gaseous SO_2, the safety regulations for liquid SO_2 are the most stringent ones.

Sulphur is an international trade product with high market potential. The world demand exceeds 35 million tons/year. The main source, about 70 % of the demand, is recovered sulphur resulting mainly from processing fossil fuels. The remaining 30 % of world's sulphur is produced by the energy consuming method of subterranean melting. Sulphur is generally stored and transported as a liquid, but intermediate stock-piling as a solid is also possible.

Sulphuric acid is a basic mass product of the chemical and related industries with a total world consumption exceeding 100 million tons/year. It is transported by rail or by sea. As sulphuric acid has the threefold transportation weight in terms of sulphur as compared to sulphur itself, it is preferably used at or close to the production site.

When comparing it to sulphur and sulphuric acid, the consumption of liquid SO_2 is very low. In Western Europe, about 300,000 tons/year of liquid SO_2 are produced, but only 50,000 to 70,000 tons/year are estimated to be sold on the market. As SO_2 is a noxious material, special regulations and precautions have to be observed for storage and handling of liquid SO_2. Land transportation is only allowed by rail.

6.2 Sodium Sulphate Byproduct

The raw sodium sulphate solution obtained in the regeneration section is preferably processed to pure sodium sulphate. The product is mainly used in the detergent, glass and textile industries. In Germany, about 200,000 to 250,000 tons year are produced, mainly from side streams of other processes. The product quality achieved at the Buschhaus plant is well ahead of the market standard, including the degree of whiteness.

There are other applications for the raw sulphate solution. For instance as make-up in the pulp and paper industry as well as for neutralizing HCl containing waste waters from other FGD plants [1].

7. Emissions and Side Streams

7.1 Treated Flue Gas

The German regulations require the reduction of the SO_2 emissions from power stations to less than 400 mg/Nm^3 (dry basis). It is anticipated that the legal emission level will be lowered to 200 mg/Nm^3 SO_2 (dry basis) at a later time. Wellman Lord plants are capable of meeting this lower level of SO_2 emissions without any plant modifications.

The chloride and fluoride emissions are reduced to less than 5 mg/Nm^3 each. The fly ash fines which have passed through the electrostatic precipitator are removed by more than 80 %. Complete SO_3 removal by ammonia injection has been applied at three Wellman Lord plants (including the Buschhaus plant). Without ammonia injection the Wellman Lord system removes SO_3 from flue gases by about 75 %.

7.2 Mother Liquor Purge

The level of impurities accumulating in the Wellman Lord solution is controlled by continuously withdrawing a defined flow of evaporator mother liquor from the circuit. In the mother liquor, the impurities are already preconcentrated.

There are two components, chloride and thiosulphate, the purge flow depends upon. The chloride content needs to be limited with respect to the stainless steel materials used in the Wellman Lord system. Futhermore there is a process requirement to carefully control the thiosulphate level as a means to minimize the mother liquor purge.

In larger scale FGD plants, the mother liquor extracted from the Wellman Lord circuit is further concentrated in a separate unit, with the effect that the impurities to be discarded are highly concentrated in the final purge, and most of the sodium sulphite contained in the mother liquor is returned to the process.

There are various options to further process or otherwise handle the final mother liquor purge:

- The liquor is sent to the prescrubber where the sodium sulphite portion is decomposed and simultaneously used for neutralizing the acidic components absorbed from the flue gas.

- The final purge is burnt in the boiler.

- The waste liquor is separately burnt in a combustion unit. The resulting residue has to be discarded.

The decision about the option finally to be chosen depends on the local conditions and especially on the environmental requirements.

7.3 Waste Water

The treated waste water contains mainly the sodium and calcium salts of HCl and H_2SO_4 formed by neutralization. The level of heavy metal ions is minimized by controlling the pH during neutralization. The precipitated solids separated from the waste water are composed of calcium fluoride, gypsum, and fly ash fines.

8. Plant Experiences

The Wellman Lord Process was presented in a number of papers published in the last two decades. These papers reflect the development and the successful application of the technology. In some of these publications, surveys were given on the operational experiences, with emphasis on the coal and petroleum coke based plants in the U.S., but they also reported about oil-based installations. Reference is especially made to the papers from A. Giovanetti et al. [2] and from U. Neumann [3] and the literature cited therein.

For the above reason, this paper describes only the new generation of Wellman Lord plants built in Europe.

8.1 The ÖMV Plant at Schwechat/Austria

At the Schwechat refinery, two power stations serve for supplying the energy required within the refinery. The FGD plant has been designed for desulphurizing the flue gases from power station 2 and off-gases from refinery units with a design flow rate of 600,000 Nm^3/hr. In power station 2, heavy residual fuels with a sulphur content of 3 to 6 % wt. were burnt. The SO_2 content in the mixed gas varied between 4 and 7 g/Nm^3.

The plant was commissioned in 1985 and operated during the first years with an averge flow rate of about 400,000 Nm^3/hr. After the flue gases from power station 1 were also added to the flue gas feed in 1987, the plant operated close to its design rate.

For cleaning the flue gases, two absorber trains for 300,000 Nm^3/h each were installed. Each train was equipped with an electrostatic precipitator (ESP) and a rotating type heat exchanger for flue gas reheating. The flue gas fan is situated behind the absorber. Sulphur trioxide is removed from the flue gas by NH_3 injection and is separated as $(NH_4)_2SO_4$ dust in the ESP's.

The regeneration unit is a conventional one with a double effect evaporator system. The recovered SO_2 gas is reduced by hydrogen in a Redotherm chamber and then converted in an existing Claus plant to liquid sulphur. The FGD plant has been designed for recovering 2940 kg/hr of SO_2.

The ESP dust removal sections caused a major problem due to the unusual consistency and properties of ammonium sulphate dust. After some modifications had been made, and experiences gained, the dust removal system has been operating satisfactorily. Some other problems were solved more easily during the early stage of commissioning. The plant has been operating smoothly and according to design, with the exception of some more sulphate formation.

More detailed information on the ÖMV plant has been published by L. Himsl and P. Reichel [4].

8.2 The BKB Plants at Buschhaus and Offleben

These Wellman Lord plants serve the Buschhaus and Offleben BLock C power stations of the Braunschweigische Kohlen-Bergwerke AG (BKB). The contract was awarded to Davy in December 1984. The plants were funded by the German Federal Government as well as by the State Government of Niedersachsen. After a very short engineering and construction period of 27 months, plant operation started in early April 1987.

The SO_2 removal capacity of both the Buschhaus and the Offleben plant with central regeneration and sulphur production at the Buschhaus site exceeds 160,000 tons/year of SO_2. These plants represent by far the largest energy-related SO_2 removal capacity on a worldwide basis. Although the BKB power stations add only about 3 % to the coal-based power generation capacity in West Germany, the share in the overall SO_2 removal capacity is in the range of 15 %.

The amount of materials that had to be handled was a major reason for BKB to choose a regenerative technology for coping with the SO_2 emission requirements. The results of a comparison between the Wellman Lord system and a limestone based technology made by BKB [5] are shown in **TABLE 3**.

Whereas the sulphur content in the Offleben lignite is "only" in the range of 1.0 to 2.5 %, the lignite burned in the Buschhaus power station contains 2 to 3.5 % wt. sulphur, with peaks up to 4 %. The SO_2 content in the flue gas is correspondingly high, ranging from 10 to 20 grams/Nm^3 (dry basis, 6 % O_2). As the SO_2 in the clean gas is limited to less than 400 mg/Nm^3, an SO_2 removal efficiency of up to 98 % is achieved.

TABLE 3

BKB's SO$_2$ Removal Project: Materials Handling Comparison
(Material flows in tons/year)

	Wellman Lord Plant	Limestone Based Plant
Main FGD Product	sulphur	gypsum
SO$_2$ in flue gas (Buschhaus and Offleben)		
– raw gas	180,000	
– clean gas	6,000 *	
Materials imported:		
– limestone		250,000
– NaOH solution (50 % wt.)	10,000	
Products:		
– gypsum		460,000
– sulphur	80,000	
– sodium sulphate	10,000	

* Start-up and shutdown periods excluded.

At each of the locations, Buschhaus and Offleben, there are two absorber trains, each designed to treat a gas flow of 800,000 Nm^3/hr. The system includes ammonia injection for SO_3 removal, dust removal by electrostatic precipitation, and reheating by liquid-linked heat exchangers.

For SO_2 absorption, packed contacting stages are used in the upper part of the absorbers. The design and structure of the absorption section is illustrated by **FIGURE 8**.

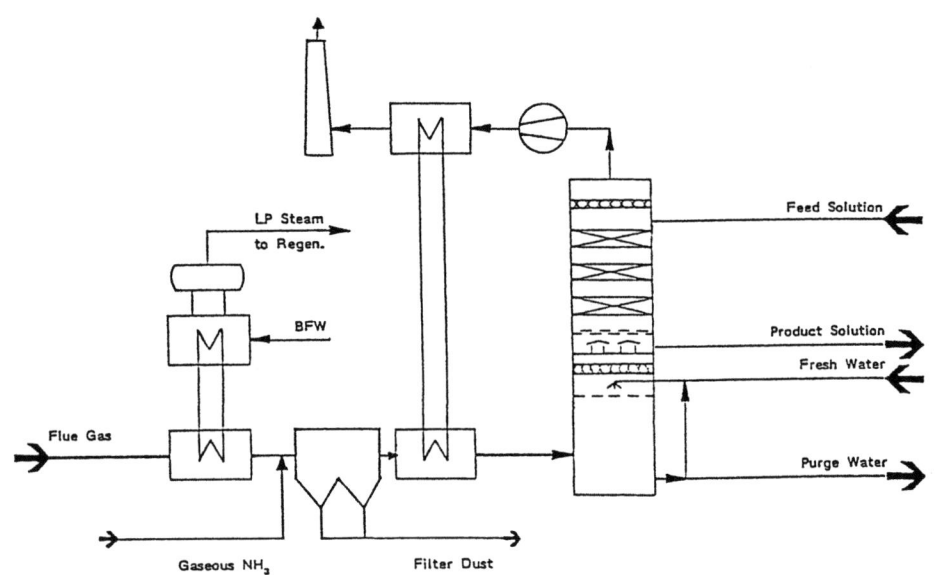

FIGURE 8

Flue Gas Treatment Section
of the Buschhaus FGD Plant

Regeneration of the Wellman Lord solution takes place in two regeneration trains with 3 evaporator crystallizers each. The regeneration plant is the first to use mechanical vapour recompression. The regeneration plant also includes a sulphate separation unit and a mother liquor concentration unit. The raw sulphate separated is further processed to pure sodium sulphate, the final mother liquor purge is burned in the boiler.

The SO_2 gas recovered in the regeneration unit is further processed to liquid sulphur. Natural gas is used for the reduction of SO_2 which takes place in an Allied catalytic reduction system followed by a two-stage Claus section. The Claus tail gas passes through a thermal incineration unit and is then returned to the absorption unit for SO_2 removal. For sulphur production, two Allied-Claus trains are used.

The single train waste water treatment unit includes ammonia recovering which takes place in a packed stripping column followed by an ammonia concentrator. The gaseous ammonia is returned to the flue gas treatment section. The treated waste water and the waste water sludge are added to the ash disposal system. The design of the unit is described in **FIGURE 9**.

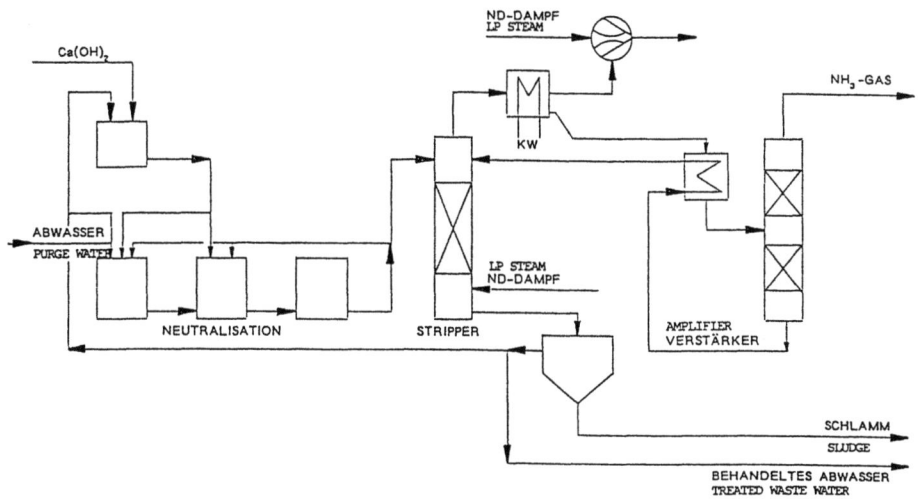

FIGURE 9

Waste Water Treatment Unit
at the Buschhaus FGD Plant

Due to the severe flue gas conditions with respect to SO_3 content and temperatures, major upsets appeared in the flue gas treatment section which required modifications and caused delays in the start-up of the FGD plant. With respect to the Wellman Lord process, a major problem resulted from a sulphate formation rate exceeding the design figure. Furthermore, the Allied unit had to be modified from a swing mode of operation to steady state operation.

A summary of the start-up problems at the BKB plants is given in **TABLE 4**. For more detailed information, see the paper of D.J. Wahl et al. [5].

TABLE 4

Start-up Problems at the Buschhaus/Offleben FGD Plants

Flue Gas Treatment Section
(Connected to Boiler)

System	Problem	Solution
Raw Gas Cooler	Mechanical Failure of Graphite Heat Exchangers	Replaced by Teflon/ Teflon Coated Heat Exchangers
SO_3 Separation	Ammonium Sulphate Tends to Clump	Various Measures: Mechanical & Process
Flue Gas Ducts	Duct Linings Damaged by SO_3 Corrosion	Replaced by Glass Flake Lining
Clean Gas Reheating	Failure of Graphite Exchanger	Replaced by Steel Based Exchangers

REGENERATION AREA

System	Problem	Solution
Evaporator Centrifuges	– Erosion on Rotor Edges – Gear Problems	Hard Metal Lining Adjusted by Vendor
Sulphate Separation and Processing	Design Capacity Insufficient	– Extension of Sulphate Train Capacity – Inhibiting Sulphate Formation
Sulphur Production	Catalyst Bed Buildup	Process Modification

8.3 The BASF Plants at Ludwigshafen and Marl

The "Kraftwerk Mitte" of BASF is located in the centre of the Ludwigshafen factory. There are 7 boilers of different sizes in operation for supplying steam and electrical power to the factory. They are all burning hard coal with low sulphur content. Dried sludge from waste water treatment or optionally other liquids or solid wastes are also being burned in one or the other boiler.

For cleaning the flue gases from power station "Mitte", two trains were installed, each designed for a flue gas flow of 570,000 Nm^3/hr. Each train includes, in the sequence as listed,

- NO_x removal by selective catalytic reduction. The installation was added later and has recently been put into operation.

- Dust removal by electrostatic precipitation.

- Flue gas cooling by the rotating system ("Drehgavo").

- Prescrubbing and SO_2 absorption.

- Flue gas fan.

- Clean gas reheating.

The absorption tower was completely manufactured from glass fiber reinforced polyester. This material of construction had proven that it was an excellent alternative to the standard Wellman-Lord design which used brick-lined carbon steel in the prescrubber section and stainless steel in the absorption section. Two packed stages were installed in the prescrubber section and four packed stages in the absorption section.

In the regeneration area a double effect evaporation system was applied: the vapours of the first evaporator were used as heating medium for the second evaporator. The section included a sulphate separation unit. This unit had to be extended as the sulphate formation rate was higher than provided for by plant design. The SO_2 recovered is processed to liquid SO_2, the mother liquor purge is burned in the fused-type boilers.

The Marl Plant is basically the same design as the Ludwigshafen plant, with the exception that the SO_2 recovery capacity is larger. Low-grade coal from the close-by BASF coal mine was burned in the two boilers of the Marl power station. The start-up of the FGD plant took place in summer 1988.

The sulphate crystallizer was sufficient in design for the separation of the sulphate formed, the raw sulphate was used at another FGD plant [1]. At the end of 1989, a mother liquor concentrator was added which operated well. The final purge was again burned in the two boilers.

Unfortunately, the Marl Power Station was taken out of operation in spring 1990, due to reasons which were not at all attributed to the FGD plant.

8.4 The Rummelsburg Plant at East Berlin

The Rummelsburg installation was engineered and built by Davy London. The contract was awarded to Davy in June 1986. he FGD plant serves a combined power and district heating station. It uses a single absorber to desulphurize the flue gases from two lignite-fired boilers, each of which can generate 320 tons/hour of steam. In the absorber train, up to 1,153,000 Nm^3/hr of gas can be treated. The design SO_2 content in the flue gas is 4,860 mg/Nm^3. Removal of SO_3 is performed by ammonia injection. For reheating of the flue gas, liquid-linked heat exchangers made of high alloy were used. In addition to the heat required for reheating, more than 20 MW of heat are recovered and added to the district heating system.

Regeneration is performed in a single evaporator equipped with mechanical vapour recompression. This unit is the largest one ever used in a Wellman Lord plant. The final product is liquid SO_2 of high purity.

The experiences gained at the Buschhaus Plant led to some modifications that were incorporated into the Rummelsburg Plant during the construction period. This does not apply to the flue gas ducts and heat exchangers. Modifications were also required in these sections which then caused delays in plant commissioning. After the FGD plant had successfully demonstrated its performance parameters, it was handed over to the end user in June 1990.

9. SUMMARY

The Wellman Lord Process represents a regenerative type technologoy for flue gas desulphurization which has been commercially experienced in the third or even the forth generation.

The technology is well proven with flue gases originated from a wide range of fossil fuels as well as off-gases from refinery and chemical plants. During the last two decades, 37 plants were built in the U.S., in Japan and in Europe with a total SO_2 removal capacity of about 850,000 tons/year.

The process recovers the sulphur dioxide removed from off-gases as a concentrated gas. The SO_2 gas is immediately converted to valuable and higly marketable sulphur products which meet the highest quality standards.

The process has been continuously improved with respect to energy savings and effluents flows. Special efforts have been made to develop new methods and paths for minimizing purge flows from the plant.

The process offers substantial benefits with respect to the logistics of an FGD project, as it consumes only low volumes of reagents and generates, as compared to gypsum processes, considerably less quantities or products.

The application of the process is favoured, when flue gases with high SO_2 concentrations have to be treated, and FGD installations with large SO_2 removal capacities are required. The drawback of higher investment cost is then significantly balanced by the benefits from the sale of the valuable sulphur products.

10. References

[1] H. Gutberlet, "Verfahren zum Aufbereiten von insbesondere aus der Rauchgasentschwefelung stammender konzentrierter Lösung von Natriumsulfit und -sulfat", DE 39 10 130 A1 (1990)

[2] A. Giovanetti, U. Neumann and R. Vangala, "Wellman Lord SO_2 Recovery Operating Experience Serving Coal Fired Boilers", Coal Technology Europe '82 Conference, Copenhagen, September 1982

[3] U. Neumann, "Das Wellman-Lord-Verfahren" in "Dokumentation Rauchgasreinigung: Betriebserfahrungen mit Rauchgasentschwefelungsanlagen", VDI-Verlag Düsseldorf, September 1985

[4] L. Himsl und P. Reichel, "Regenerative Rauchgasentschwefelung nach dem Wellman-Lord-Verfahren in einem Industriekraftwerk", VGB Kraftwerkstechnik 88, 56 - 59 (1988)

[5] D.-J. Wahl, J. Rahm und H.-H. Grimm, "Erfahrungen mit der Rauchgasentschwefelung nach dem Wellman-Lord-Verfahren", Kraftwerk und Umwelt 1989, 249 - 253

[6] D.-J. Wahl, "Rauchgasentschwefelung der Braunkohlenkraftwerke Buschhaus und Offleben C" in "Sonderteil Betriebserfahrungen REA", Brennstoff-Wärme-Kraft 3, L3 - L10 (1990)

ISPRA MARK 13A DESULPHURIZATION PROCESS

H. LANGENKAMP
Commision of the European Communities,
Joint Research Centre,
Environment Institute
I-21020 ISPRA (Varsese), Italy

1. Introduction

The most common method actually used for flue gas desulphurization is the wet method, consisting of lime or limestone scrubbing. This method, although almost universally applied, has some intrinsic drawbacks: costs are incurred in handling the lime and limestone and the resulting large quantities of wet slurries, considerable amounts of waste water are produced and the disposal of large quantities of gypsum will cause problems.

For these reasons, new regenerative processes may still offer considerable advantages.

2. Process History

The basic idea for the ISPRA MARK 13A process was generated in the Joint Research Centre of the European Communities at Ispra (Italy) in 1979 as a spin-off of the hydrogen energy research programme of the European Community [1].
Research on the thermochemical decomposition of water into hydrogen and oxygen was started in 1970 in several places of the world. In 1974, a variant of the purely thermochemical cycles was introduced in the hybrid cycle. Here electrical energy is used in one reaction of the process. One example of potentially promising hybrid cycles is the ISPRA MARK 13 (13 = process development numbers) sulfur-bromine cycle [1]. The cycle consists of the following three reactions:

(1) $SO_2 + Br_2 + 2 H_2O$ → $2 HBr + H_2SO_4$
(2) $2 HBr$ (electrochemical step) → $H_2 + Br_2$
(3) H_2SO_4 (850 °C) → $H_2O + SO_2 + \frac{1}{2} O_2$

A continuous laboratory-scale plant of the process was constructed and operated at the JRC Ispra. The nominal hydrogen production rate of the plant was 100 l/h. The separation of SO_2 and O_2 was performed primarily by cooling down the gas mixture to -48°C. The gas mixture, leaving the cooling trap consisted of 93-95% O_2 and 5-7% vol. SO_2.
The remaining SO_2 was removed by contacting the gas with a diluted bromine water solution, the sulphur dioxide was oxidised and SO_2-free oxygen leaves the

process. This purification step was the basic idea for the creation of the new desulphurisation process.

2.1 The ISPRA MARK 13A desulphurisation process.

The new process ISPRA MARK 13A was invented in 1979 and the first tests with a air/SO_2 gas mixture were started in 1981 [2]. In 1981/82 a preliminary feasibility study was carried out by Prof Gestrich of the T.U. Berlin [3].

Bench scale experiments with flue gas connected to an oil fired power station in Ispra were carried out in 1982. At the end of 1982 the construction of a transportable unit for testing the process on a coal fired furnace in Livorno was started. In the meantime we started also the determination of physical properties (vapour pressure of liquid mixtures of HBr/H_2SO_4) and did other supporting work. Between 1983 and 1984 tests on a coal-fired experimental furnace were done at the laboratories of ENEL in Livorno.

In the same period, the firm COMPRIMO in Amsterdam carried out a preliminary engineering design and cost evaluation study for a pilot plant of 20.000 Nm^3/h [4] and the firm UHDE in Dortmund prepared a detailed study for an electrolyser for this plant [5].

Both studies served as a basis for the ISPRA MARK 13A Pilot Plant Project, which was started in 1985 with the call for proposals and negociations with the different candidate firms. In 1986 a contract was signed for the construction of a pilot plant of 32.000 Nm^3/h with the firm FERLINI TECHNOLOGY. The engineering, construction and start up was subcontracted to KRAFTANLAGEN Heidelberg. The site of construction in the SARAS Refinery in Sarroch, Sardinia. Construction started in 1987 and the plant was started up at the beginning of 1989. The test programme will be finished at the end of 1991.

3. Process description

The ISPRA MARK 13A prcess is a wet scrubbing regenerative process, producing two valuable chemicals, sulphuric acid and hydrogen. It does not generate any waste water. A simplified diagram of the process principle is given in **Fig. 1**.

The process is based on the following two reactions:

(1) $SO_2 + Br_2 + 2 H_2O \rightarrow HBr + H_2SO_4$
(2) $2HBr$ (electrolysis) $\rightarrow H_2 + Br_2$
Sum $SO_2 + 2 H_2O \rightarrow H_2 + H_2SO_4$

The reactive agent is a dilute aqueous solution of 10-20 wt% H_2SO_4 and HBr containing **a small amount of bromine** (<0,5 wt%). The SO_2 containing flue gases are brought in contact with this solution in the reactor. Suphur dioxide is absorbed in the solution and reacts immediately in the liquid phase according to reaction 1).

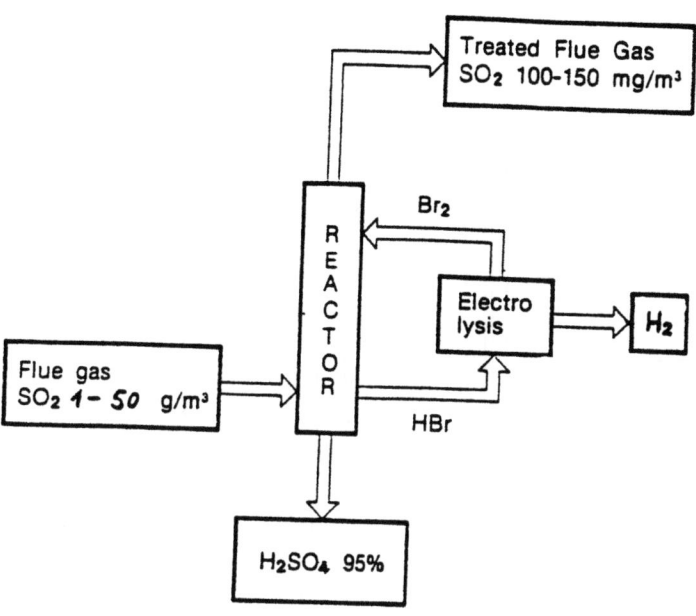

Fig. 1
Simplified Flow Diagram

In this reaction the bromine conversion is 100%, a bromine free washing liquid is produced and the reaction products, hydrogen bromide and sulphuric acid, stay dissolved in the solution. A block scheme is given in Fig. 2.

The bromine consumed in reaction 1 **is regenerated by electrolysis.** To this end, the bromine free washing liquid is passed through an electrolytic cell, where the original Br_2-content is re-established and hydrogen is formed simultaneously (reaction 2). It follows that the active reagent of the process, bromine, is completely recycled and hence does not contaminate the environment. It has to be stressed that in the process bromine occurs only in very dilute solutions (< 0,5 wt%). In these concentration this chemical can be considered harmless. One of the main advantages of the process is that no chemicals have to be added. In most other FGD processes, the consumption of absorption chemicals constitutes a major item of the operation costs.

The three final products of the ISPRA MARK 13A process are hydrogen, sulphuric acid and desulphurized flue gas. The first one, **hydrogen**, is directly generated in the electrolytic cell in a high purity. Only traces of hydrogen bromide and bromine vapour have to be removed, which occurs by washing with a suitable liquid, e.g. reactor solution. Hydrogen has many attractive applications in the chemical industry and it can also be used as a co-fuel in the power station.

The other end product, **sulphuric acid,** is primarily produced in the reactor solution from which it has to be concentrated and separated. This separation is

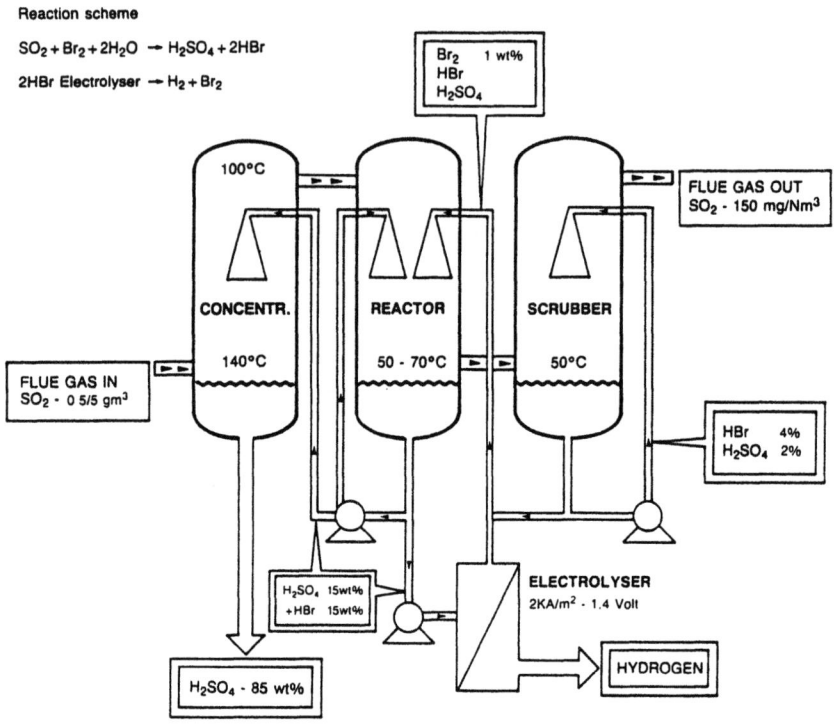

Fig. 2
Block scheme of the Ispra Mark 13A FGD process

accomplished by evaporation, making use of the sensible heat of the incoming flue gas (150-160 °C). A side stream of the reactor liquid is fed to the pre-concentrator where it flows countercurrent to a part of the entering flue gas. Here all HBr and a large part of the water from the reactor liquid are evaporated and 70 wt% H_2SO_4 solution is produced.

The acid can be further concentrated to 95 wt% in the final concentrator. For this operation a small stream of hot gas of about 300°C is required. The hot gas stream can be taken from a suitable stream in the boiler unit (preheated air or hot flue gas).

The **desulphurized flue gas,** after leaving the reactor and the demister, still contains some reactor liquid and HBr vapour. Therefore a final scrubber is placed downstream of the reactor where the contaminants are washed out with water and returned to the reactor fluid. The water inventory of the process is maintained either by condensing water from the flue gas or by addition of demineralised water. Anyhow, no waste water stream is produced.

The most innovative part of the ISPRA MARK 13A process is the HBr electrolysis of mixed HBr 15% and H_2SO_4 15 wt% electrolyte. The produced bromine in the electrolyser remains in solution and leaves the electrolyte as a mixture of H_2SO_4/HBr/Br_2 15/14.5/0.5 wt%.

The bromine produced on the electrolyser is directly proportional to the amount of SO_2 removed from the fluegas.

For safety reasons is has to be guaranteed that always 10-30 mg/m^3 of SO_2 will be left in the treated fluegas.
The production of the active reagent (Br_2) by electrolysis avoids the addition of chemicals to the process, i.e. consumption of chemicals, storage facilities etc. However, this advantage is counterbalanced by an increase of the power consumption of the process.

4. Pilot Plant

After the successful operation of the bench-scale units for oil and coal fired boilers, the next logical step in the development of the process was the construction and operation of a demonstration pilot plant.

The ISPRA MARK 13A Pilot Plant Project was organised in collaboration by the JRC Ispra and the service "Valorisation of Research Results" (DG XIII) of the European Communities in Luxemburg. It had the form of an open call for proposals published in the Official Journal of the European Communities in November 1984 which invited proposals for the construction and operation of a suitable pilot plant.
The pilot plant has been built by the Firm FERLINI TECHNOLOGY in Sarroch near Cagliari (Sardinia) at the SARAS Refinery.
SARAS is one of the largest refineries in the mediterranean area with a total throughput of 18 million tons of crude oil per year.

The plant is designed for a throughput of 32 000 m^3/h of flue gas produced by the combustion of a mixture of heavy fuel oil and refinery gas, preliminary purified by an electrostatic precipitator. Flue gases with an average content of 0.16 vol% SO_2 (4500 mg/m^3) are treated. The degree of desulphurization should be higher than 90%. The sulphuric acid produced is used in the refinery for the regeneration of ion-exchange beds for boiler feed water preparation.

A block scheme of the ISPRA MARK 13A pilot plant in the SARAS Refinery is given in **Fig. 3**.

4.1. Acid formation Section

The formation of sulphuric acid and hydrobromic acid takes place in the reactor of the plant. (**Fig. 3**)
The reactor is an open spray column of a diameter of 2 m, and a column height of 13 m. There are 20 spray nozzles, 4 nozzles for each level installed. The operating temperature is approximately 65°C. The reactor has an internal recycle of 120 m^3/h of a dilute acid solution of $\approx$ 15 wt % HBr / $\approx$ 15 wt% H_2SO_4. Another stream of this acid solution containing $\approx$ 0.5 wt% of Br_2 is coming from the electrolyser section. The bromine of this solution reacts with the SO_2 in the fluegas to H_2SO_4 and HBr. The reactor operates in the co-current mode: Flue gas and the acid bromine solution are both entering the top of the reactor. All bromine reacts and a bromine free gas stream leaves the reactor at the bottom end via a demister for droplet separation.
The desulphurized flue gas still contains some traces of reactor liquid and HBr vapour. Therefore, a final scrubber is placed downstream of the reactor where

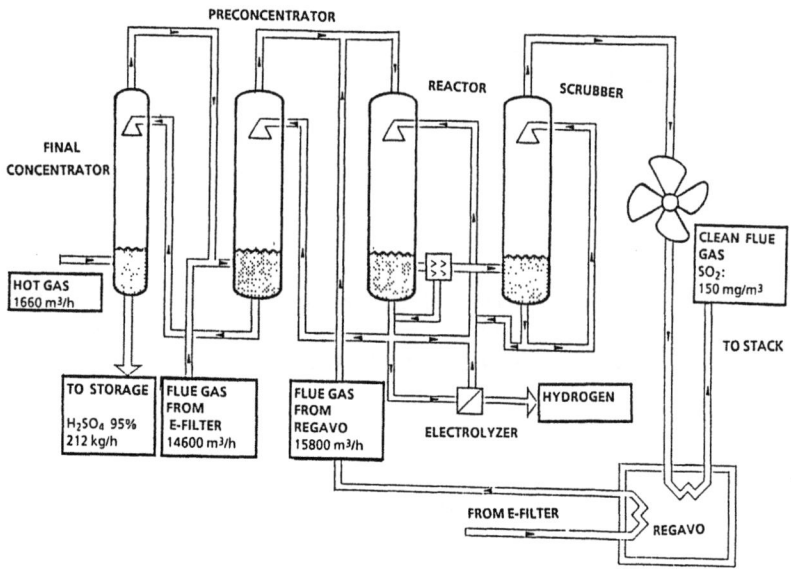

Fig. 3
Block scheme of the Ispra Mark 13A pilot plant

the contaminants are washed out with water The produced scrubber liquid contains 0.2 wt% HBr and 0.2 wt% H_2SO_4 and is afterwards returned to the reactor fluid. The operating temperature of the scrubber is 45 - 50°C.

The water inventory of the process is maintained either by condensing water from the fluegas or by addition of water to the scrubber. The desulphurized flue gas passes through a demister for droplets separation and is reheated in a Lungstroem rotating heat exchanger (GAVO) from 50°C to 125°C before entering the stack.

4.2. The Electrolysis Section.

Hydrobromic acid electrolysis has been studied during the hydrogen energy programme about 1980, but mainly in concentrated solutions (47.5 wt%) and with graphite electrodes. The application in mixed hydrobromic/sulphuric acid acid solutions is new and the use of metallic electrodes has not been reported. In the present development, the existing experience with the electrolysis of hydrochloric acid forms the basis of the cell design.

The HBr electrolysis in the pilot plant on Sarroch is carried out in two parallel cell units from different companies and of different design. Both electrolysis units are working without membrane or diaphragm.

The maximum bromine concentration on the electrolyte can be varied between 0.2-0.5 wt%. The bromine production rate needed for a 90% desulphurization is 320 kg/h. This requires an electrolysis current of about 110 kA. The individual cell

voltage is usually between 1.3 and 1.6 V, so that the power consumption for electrolysis is approximately 150 kWe, i.e. 1.2% of the total plant output.
Here it must be taken into account that the calorific value of the hydrogen produced in the electrolysis (50 Nm^3/h) amounts to 170 KW, which for a large part compensates the electrolysis power consumption.

4.2.1 The DEM Cell

The electrochemical cell used for the laboratory tests in Ispra in preparation for the pilot plant was developed by the Electricity Council Research Centre [Capenhurst, Chester, England] and is manufactured under license by Electrocatalytic Ltd. It is known as the "Dished Electrode Membrane" cell (DEM-cell). A cross-section of the cell is shown in **Fig. 4**.

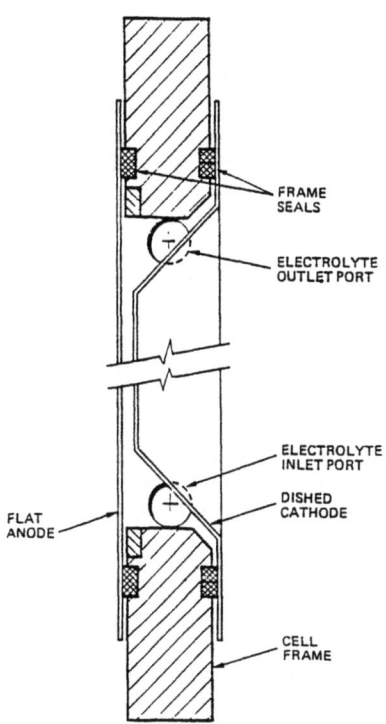

Fig. 4
Cross section of an undivided DEM cell

It is an undivided monopolar electrolyser fitted with metallic electrodes which consist of a dished cathode (Hastelloy C) and a flat coated anode. The interelectrode gap is small, 4-5 mm, thereby minimizing the internal resistance of the cell, and hence the cell voltage. The dishing of the electrode allows a comparatively thick cell frame to be used, which accomodates large electrolyte ports for connections to external manifolds. High electrolyte flow rates can easily be accomodated which is essential for high mass transport conditions. The use of heavy gauge aluminium current distribution plates in the DEM cell provides a good current distribution across the face of the electrode.

As the use of metallic electrodes is a novelty in the electrochemical decomposition of hydrobromic acid, a careful control of the corrosion rate of both electrodes has been carried out. It must be noted that the DEM cell has already found a number of industrial applications, but had never been tested for bromine production at such low pH.

The anode material must preferably show a low overpotential for Br_2 evolution and a high O_2 overpotential. A dimensionally stable anode (DSA) consisting of a mixed RuO_2/TiO_2 coating on a flat titanium sheet was chosen. This material is industrially used for chlorine electrolysis, where the chlorine overpotential is as low as 5 to 40 mV [6]. The oxygen overpotential on this electrode material has been determined by Scarpellino [7] and is approximately 200 mV.

As far as the cathode is concerned, a low H_2 overpotential and a good resistance in the corrosive electrolyte are required. The selected cathode material was Hastelloy C 276. The main constituent of this alloy is Ni.

The maximum amount of bromine to be produced by the electrolytic cell is directly proportional to the incoming SO_2. The plant has a capacity of 32.000 Nm^3/h of flue gas with a SO_2 content of about 4500 mg/m^3. This implies a total flow rate of 144 kg/h SO_2, which requires at 90% desulphurization an equivalent quality of 338 kg/h of bromine.

The flow rate of electrolyte through the electrolyser can be varied between 30-50 m^3/h. The temperature of the electrolyser is between 60-70 °C.

From our laboratory experiment we expect a cell voltage in the range 1.4 - 1.6 V at ≈ 2 kA/m^2. A current/voltage curve at 50°C is presented in **Fig. 5** and the influence of temperature on the cell voltage is given in **Fig.6**.

Finally it must be said that the final results from the pilot plant DEM cell electrolyser are as yet outstanding. Work is under way and the results of a series of long duration are expected to be available at the end of 1991.

<u>4.2.2. The Deutsche Carbone Electrolyser</u> This electrolyser was developed by the Deursche Carbone Aktion Gesellschaft (DCAG) on the basis of the existing Envirocell-system, an electrochemical waste water treatment process.

Both electrodes, cathode and anode are made of graphite, whereas the frames and the cell housing are made in PVDF (polyvinylidene fluoride). The electrodes are vertical with a gap of 10 mm. Each electrode is 40 mm thick. There are no separators, membranes and turbulence promoters between the electrodes. The electrolyte flows over a distribution plate in the bottom of the electrolyser. The hydrogen leaves the electrolyser together with the electolyte and is separated in a degasifier. An illustration of the DCAG electrolyser is given in **Fig. 7**.

The total cell surface area of the cell is 60 m^2. There are three cells in parallel each with 41 electrodes. From the start up on the pilot plant up till now, only the DCAG electrolyser has been in operation. The total running time of the electrolyser up to August 1990 was 550 h, corresponding to a total H_2SO_4 production of 85 ton.

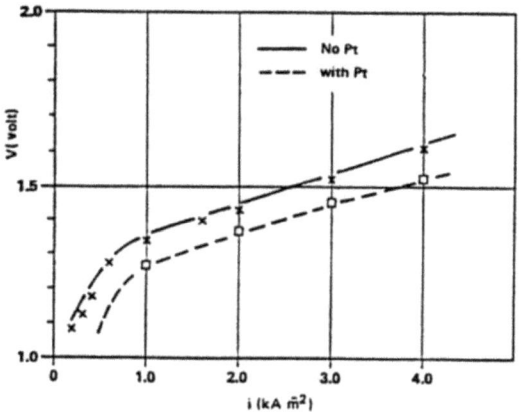
Fig. 5 Current/voltage curve at 50°C.

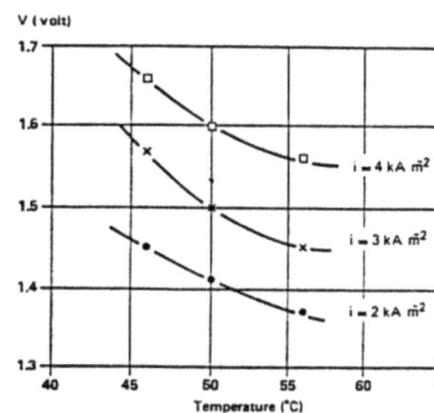
Fig. 6 The influence of temperature on cell voltage.

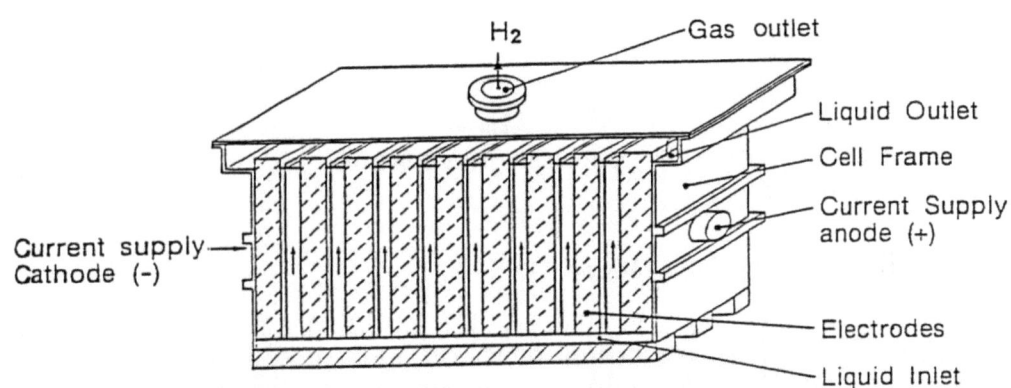
Fig.7
DCAG Electrolyser (Enviro-cell type)

The current efficiency obtained with a laboratory electrolyser from DCAG (electrode surface 1800 cm2 = 20 x 90 cm) in the Ispra bench scale plant, was about 70% at an electrolyte flow rate of 300 kg/h. The electrolyte consisted of HBr 18 wt%/ H_2SO_4 22 wt% .The operating temperature was 45°C.
In the pilot plant DCAG electrolyser, the electrolyte flow rate was 28 m3/h, the electrolyte consisted of 13.3 wt% HBr and 11,7 wt% of H_2SO_4 and the

temperature was 60°C. Under these conditions, the current efficiency remained very low. Only values of 30-50 % could be reached

Further modifications and experiments are necessary to bring the current efficiency of the DCAG cell to an acceptable level of 60-70%. The current/voltage curves from laboratory and pilot plant electrolysers are presented in **Fig. 8 and 9**.

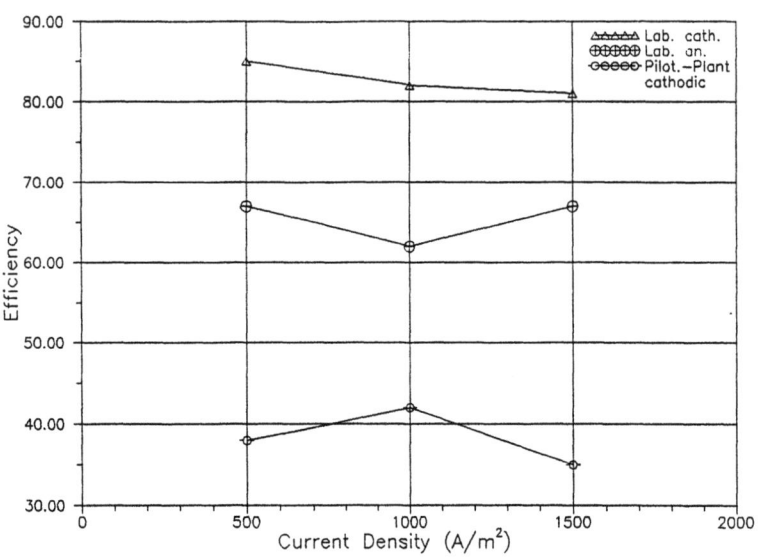

Fig. 8
Current efficiency DCAG cell

4.3. Acid Concentration Section

4.3.1. Pre-concentrator The sulphuric acid produced in the reactor solution has to be separated from the hydrobromic acid and concentrated. This separation step is accomplished by evaporation, making use of the sensible heat of the incoming flue gas (160-170°C).
A side stream of the reactor liquid is fed continuously to the pre-concentrator, where it is contacted countercurrently with the incoming hot flue gas. The liquid enters the column via a set of spray nozzles. The column is an open spray tower of 2.0 m internal diameter and a height of 8.5 m.
In the preconcentration step all HBr and a large part of the water from the reactor liquid is evaporated and a 65-70% wt% H_2SO_4 solution is produced. The flue gas leaves the preconcentrator with a temperature of 110 °C, the liquid temperature in the preconcentrator being 70-75 °C.
During pilot plant experiments the HBr concentration of the produced H_2SO_4 was less than 0.2 wt%. The produced sulphuric acid has a light-brown colour.
A chemical analysis of the produced acid is given in **Table 1**.

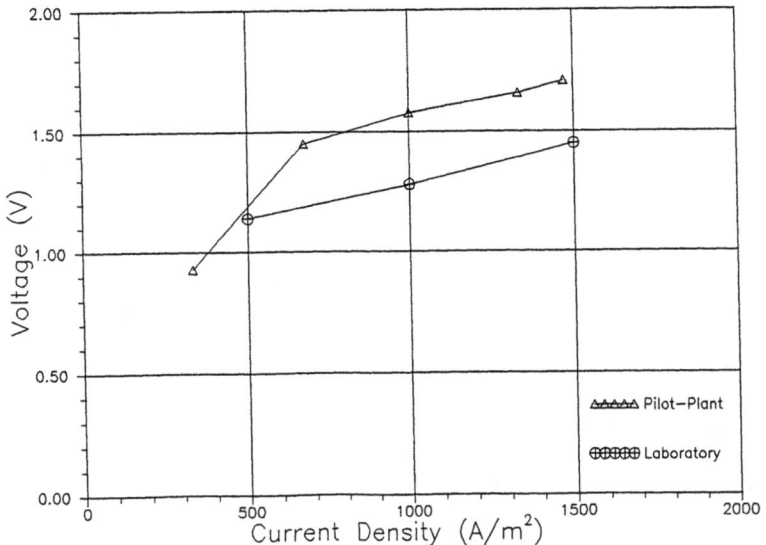

Fig. 9
Current/voltage curves DCAG cell

TABLE 1

Sulphuric acid composition from ISPRA MARK 13A plant in Sarroch Preconcentrator			
Density:		1.585	g.ml^{-1}
H_2SO_4 content		67.6	wt%
HBr content		0.1	wt%
Residue at 300 - 350°C		0.22	wt%
Main impurities:			
Na		500	ppm
Fe		317	ppm
Ca		300	ppm
SiO_2		n.d.	
Ti		210	ppm
Mg		40	ppm
Al		34	ppm
Cr		27	ppm
Mn		16	ppm
V		15	ppm
Zn		14	ppm
Cd	less than	1	ppm
All values refer to original solution			

4.3.2 Final concentrator

The sulphuric acid of ~70 wt% produced in the pre-concentrator can be further concentrated to 95 wt%. Various solutions to this problem are available:

a) By direct gas contact
   When a suitable hot gas stream, of more than 300°C is available, the final concentration step can be carried out by direct liquid/gas contact. This is the case in the Sarroch pilot plant, where hot flue gas coming from an incinerator at 350-400°C can be used for this purpose.
   The final concentrator is a tower of 2 m internal diameter with 1 m of ceramic packing material. The column is operating counter-current, sulphuric acid from the pre-concentrator is fed to the top of the column by a set of spray nozzles and the hot gas is entering at the bottom of the column. Final results of the tests with the final concentrator in Sarroch are as yet oustanding.

b) By classical concentration processes
   If a hot gas stream is not available, the final concentration can be carried out by current acid concentration processes. A good example is the Bertrams' process which is a simple and reliable process. The process consists of the evaporation of the excess of water in a forced circulation loop under a moderate vacuum of 60-90 mbar. In case of the presence of organic impurities, oxidation agents are added. More than ten reference plants of the process are actually in operation. A sketch of the Bertrams' process is given in **Fig. 10.**

The firm Bertrams disposes of a pilot plant installation in Muttenz (Basel, Switzerland). This plant has a capacity of about 100 kg/h of 95 wt% H_2SO_4. In this plant an experimental run was carried out where 2500 kg of preconcentrator acid from the Sarroch pilot plant was concentrated from 67.5 to 95 wt%.
The results obtained from this experiment were very satisfactory.
The product acid has a concentration of more than 95 wt % and is light brown. The colour of the product acid is caused by the presence of small amounts of organic compounds. The COD value of the product acid is 535 mg/l. The organic compounds can be oxidised by a treatment of the acid with an oxidant, e.g. nitric acid of hydrogen peroxide.The solid content in the acid is low and can be eliminated by filtration or deposition. The composition of the product acid is given in Table 2.

4.4. Process control

One of the advantages of the ISPRA MARK 13A process is the simplicity of the main process control. The main parameter to be controlled is the required production and flow rate of bromine. The fact that this reagent is produced in situ by electrolysis allows the application of a very simple process control system. Two SO_2 analysers are used for this purpose, one at the inlet of the FGD process and one in the desulphurised gas. The inlet SO_2 concentration in combination with the flue gas flow rate determines the set point of the power supply to the elctrolyser. The current consumed in the electrolyser is directly proportional to the bromine production.

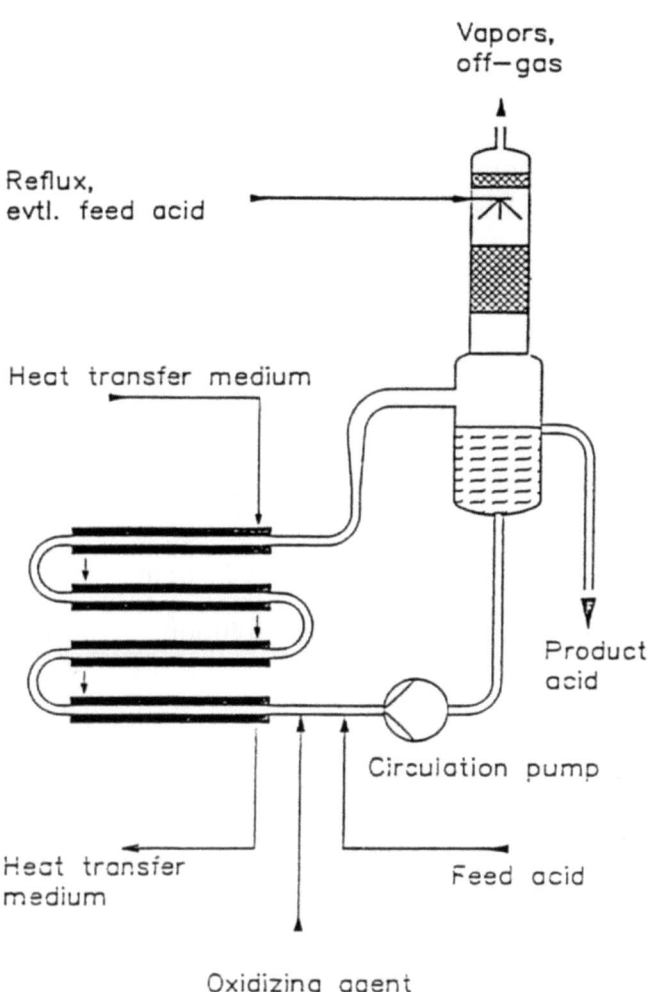

Fig. 10
Bertrams' forced circulation evaporation and purification system

The second analyser measures the SO_2 outlet concentration and is used for the fine regulation of the current. The outlet concentration of SO_2 in the fluegas will be maintained constant at a value of 30 mg/m³. A sketch of the process regulation principle is given in Fig. 11.

TABLE 2

Sulphuric acid composition from ISPRA MARK 13A plant in Sarroch Final concentration by Bertrams' process.

Density:	1.835	g.ml^{-1}
H_2SO_4 content	95.7	wt%
HBr content less than	000.1	wt%
Residue at 300 - 350°C	0.13	wt%

Main impurities:

Na	176	ppm
Fe	36	ppm
Ca	45	ppm
Ti	0.6	ppm
Mg	20	ppm
Al	8	ppm
Cr	18	ppm
Mn	16	ppm
V	3	ppm
Zn	6	ppm
Cd	0.02	ppm

All values refer to original solution

5. Materials

One of the most important items of the pilot plant operation is the testing of the applied construction materials. All liquids which are used in the process are strong acids, varying in concentration from less than 1 wt% (scrubber liquid) via approximately 30 wt% (reactor liquid and electrolysis) to high concentrations of more than 70 wt% in the concentration section.

The two scrubbing columns (reactor and scrubber) are made from glass lined polyvinylester resin [Atlac ICI 580-05]. During the life of the pilot plant, no corrosion. problems were identified with these items. The piping for this section is also made of glass lined polyvinylester. Material problems were encountered in the flanges by imperfect glueing and, consequently, penetration of HBr. The problems were solved by replacement of all flanges and application of a protective internal layer of 2 mm of pure vinylester.

The operation conditions of the pre- and final concentrator are more severe. These pieces of equipment are therefore made of glass lined steel.
For various piping GRP, PVDF and PTFE, PVC-C (sometimes inliner) are used. When higher concentrations of sulphuric acid occur, piping is always made of steel with a PTFE inliner.

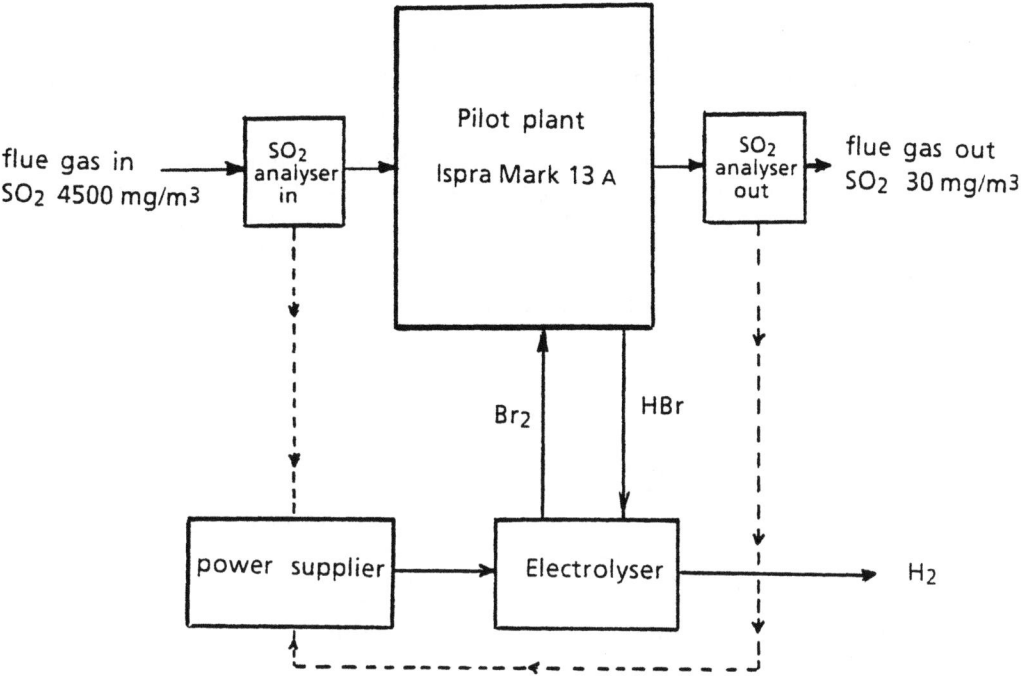

Fig. 11
Process control principle

6. Projet organisation and financing

The main features of the project organisation of the ISPRA MARK 13A Pilot Plant Project are summarised below:

Patent owner:	European Community
Licensee:	Ferlini Technology s.r.l. Brescia (Italy)
Planning, construction and start-up:	Kraftanlagen Heidelberg Heidelberg (Germany)
Plant site:	SARAS Refinery Sarroch, Sardinia (Italy)
Financing:	Participation of the European Community for 50% (5 mio ECU)
Estimated total costs, including 2 years' operation:	10 mio ECU.

7. Economics

The general tendency of the economics for regenerative FGD processes is that the investment costs are somewhat higher than those of the lime/limestone scrubbing processes and that the main advantages are in the variable operating costs.

This statement is particularly true for the ISPRA MARK 13A process, where no consumption of absorbant reagents is required at all. The electrolysis represents a certain burden to the investment costs, but the variable operating costs compare very favourably with those for the lime/limestone process. A cost comparison on the basis of the data for the pilot plant is given below, in Table 3.

A detailed cost comparison study is presented elesewhere in the present course [7].

TABLE 3

Variable costs comparison			
	Basis: Flue gas flow rate: SO_2 content	Pilot plant SARAS 32 000 Nm³/h 4.5 g/m³	
Lime/limestone scrubbing		**ISPRA MARK 13A**	
$CaCO_3$ Costs Consumption Total costs	 20 ECU/ton 240 kg/h Ca 4.80 ECU/h	*Electrolysis current* Costs Consumption Total costs	 0.072 ECU/kWh 145 kW 10.87 ECU/h
		HBr losses Costs Consumption Total costs	 2.67 ECU/kg 3 mg/m³ of flue gas 0.26 ECU/h
Credits -----		*Credits* *Sulphuric acid* Price Production Total	 45 ECU/ton 216 kg/h 9.72 ECU/h
		Hydrogen Price Production Total	 0.05 ECU/Nm³ 46 Nm³/h 2.30 ECU/h
TOTAL Costs Credits	 4.80 ECU/h ----	Costs Credits	11.13 ECU/h 12.02 ECU/h

8. Summary

- ISPRA MARK 13A is a regenerative flue gas desulphurization process without waste streams, producing two useful products: hydrogen and sulphuric acid.

- The process is entirely gas-liquid without solids handling.

- Capital costs are competitive with conventional lime-limestone processes and operating costs are substantially lower.

- In the process SO_2 is directly oxidized in a spray tower by a very dilute (%) solution of bromine, forming sulphuric acid.

- Bromine is continuously regenerated by electrolysis and high purity hydrogen is produced.

- The speed and effectiveness of the chemical reactions allows for relatively small absorption towers.

- All unit operations of the process have been tested and demonstrated in a series of test units.

- A pilot plant is presently in operation at a refinery in Italy and has demonstrated the validity of the overall process: SO_2 removal efficiency higher than 90% and high quality of the hydrogen and good quality of sulphuric acid produced.

- Materials and equipment are mostly plastic and glass lined steel.

- The process needs still to be demonstrated in a coal-fired utility on an industrial scale

9. References

[1] D. van Velzen, H. Langenkamp and G.F De Beni, " Process for the removal of SO_2 from waste gases, producing hydrogen and sulphuric acid"; Eur. patent n° 0.016290

[2] W. Gestrich, "Durchführbarkeits Studie Mark 13A-Prozess zur Abscheidung von SO_2 aus Rauchgasen mit Wasserstoff und Schwefelsäureproduktion als Abfallprodukt",Berlin, January 1982

[3] COMPRIMO B.V.,"Pilot plant Mark 13A process for flue gas desulphurisation", Final Report of Contract No. 2195-83-10 ED.IS.NL (JRC Ispra/Coomprimo B.V. Amsterdam), April 1984

[4] UHDE GmbH, "Study for electrolyser of the Mark 13A process", Final Report of Contract No. 2208-82-11 PC.ISP.D (JRC Uhde GmbH, Dortmund), May 1984

[5] D. Pletcher in "Industrial Electrochemistry", Chapman and Hall Ltd., (London) 1984. Ch. 3, 88-113

[6] A. J. Scarpellino and G. L. Fisher, J. Electrochem. Soc., 129 (1982) 515-522

[7] D. van Velzen, "Costs of desulphurisation and denoxing", Eurocourse: Sulphur dioxide and nitrogen oxides in industrial waste gases: emission, legislation and abatement", Ispra, 3-7 September 1990

BF/UHDE/MITSUI-ACTIVE COKE PROCESS FOR SIMULTANEOUS SO_2- AND NO_x-REMOVAL

E. RICHTER
DMT-GESELLSCHAFT FÜR FORSCHUNG UND PRÜFUNG MBH
INSTITUT FÜR WÄRME- UND STROMERZEUGUNG
FRANZ-FISCHER-WEG 61
4300 ESSEN 13
FED. REP. OF GERMANY

1. Introduction

The processes Bergbau-Forschung GmbH, now DMT-Gesellschaft für Forschung und Prüfung mbH, Essen (D) developed and, together with its licencees, Uhde GmbH, Dortmund (D), Mitsui Mining Company, Tokyo (J), Kinetics Technology International B.V., Zoetermeer (NL), and Steag AG, Essen (D) improved are suitable for SO_2-, SO_2/NO_x and NO_x-removal. In every case, active coke is used as adsorbent and catalyst at temperatures between 100 and 170 °C.
Active coke is a formed material produced from bituminous coal. Active coke is produced since over 25 years by a process developed by Bergwerksverband GmbH, Essen (D). The producer is Carbo-Tech GmbH, Essen (D).
Bituminous coal of low ash content is ground to very fine powder. This powder is oxidized by air in a cascade of fluidized beds at a temperature under 300 °C. The oxidized coal is mixed with pitch and water which serve as binder. The mixture is subsequently extruded so that cylindrical, so-called "green" noodles are formed. The diameters range between 2 and 9 mm. The noodles are calcined in an inert atmosphere in a rotating kiln. The calcination temperature is above 550 °C.
By calcination, micropores are formed in the initial coal fines, macro pores are formed by the decomposition of the binder.
Immediate and elementary analysis and ash compositon of active coke VA5 are as follows:

Immediate analysis
Water	(% by wt., wf)	<	3
Ash	(% by wt., wf)		2.4
Volatile matter	(% by wt., wf)	<	3

Elementary analysis
C	(% by wt., wf)	93.6
H	(% by wt., wf)	0.8
O	(% by wt., wf)	1.3
N	(% by wt., wf)	1.1
S	(% by wt., wf)	0.8

Ash analysis (oxides)

SiO_2	(% by wt.)	37	- 42
Fe_2O_3	(% by wt.)	17	- 20
Al_2O_3	(% by wt.)	27	- 30
CaO	(% by wt.)	2.0	- 2.6
MgO	(% by wt.)	1.4	- 1.7
$Na_2O + K_2O$	(% by wt.)	4.8	- 5.8

Active coke has a very low ash content so that its adsorptive and catalytic properties are mainly determined by the carbon properties. It is necessary to use a bituminous coal with low ash content because of the carbon consumption during the process (see Chapter 2.2).

The characteristics of the active coke VA5 are:
- high adsorptive capacity for SO_2,
- rapid adsorption kinetics,
- high catalytic activity for ammonia based NO_x-reduction, even at low temperatures,
- low reactivity with oxygen (ignition point over 380 °C),
- improved activity due to regeneration,
- high mechanical strength,
- low price compared to activated carbons.

The pore volume of active coke, used for SO_2-removal, grows with the progression of adsorption/regeneration cycles. This is the reason why the adsorptive properties are improved by regeneration [1].

2. Fundamentals of the Processes for SO_2- and NO_x-Removal

2.1 SO_2-Adsorption and NO_x-Reduction

SO_2 will be adsorbed on active coke over a wide temperature range, from ambient temperature to 150 °C, where it reacts with oxygen and steam to yield sulphuric acid [2].

(1) $$SO_2 + 1/2\ O_2 + H_2O \longrightarrow (H_2SO_4)_{ads}$$

Sulphuric acid is retained by the pore system of active coke. Maximum loads between 10 and 20 % by wt. are obtained even at tempertures of not higher than 120 °C as sulphuric acid has an extremely low vapour pressure [1].

In the flue gases from coal firings, nitric oxides (NO_x) are composed mainly by nitric oxide (NO) and only by 5 to 10 % of nitric oxide (NO_2). The latter is reduced quite fast by reacting with the carbon, at temperatures above 80 °C. When adding ammonia to the flue gases, the NO is reduced catalytically by active coke, as follows:

(2) $$4\ NO + 4\ NH_3 + O_2 \longrightarrow 4\ N_2 + 6\ H_2O$$

Above 80 °C any reaction between NO and NH_3 to give ammonium nitrite will be excluded as the latter substance is unstable within that temperature range. If the flue gases contain some sulphur dioxide, even after a preceding desulphurization step, sulphuric acid is formed on active coke. The acid is neutralized by ammonia according to:

(3) $\quad 2\,NH_3 + H_2SO_4 \longrightarrow (NH_4)_2SO_4$

The kinetics are described in detail in [2].

2.2 Regeneration of Active Coke

The active coke, loaded by sulphuric acid and relevant ammonium salts, needs thermal regeneration. From about 300 °C onward sulphuric acid will react with the carbon of the active coke to sulphur dioxide and carbonaceous surface oxides C...O as an intermediate product, as per following expressions:

(4) $\quad 2\,H_2SO_4 + 2\,C \longrightarrow 2\,SO_2 + 2\,H_2O + 2\,C...O$

(5) $\quad 2\,C...O \longrightarrow C + CO_2$

Ammonium sulfate decomposes (opposed to reaction (3)) to form sulphuric acid and ammonia. The latter will spontaneously reduce SO_3 (intermediate product of the degradation of sulphuric acid) or surface oxides as per

$$2\,NH_3 + 3\,C...O \longrightarrow N_2 + 3\,H_2O + 3\,C$$

This is why carbon losses are minimized during regeneration.
The mechanisms and kinetics of thermal desorption of loaded active coke are described in detail by [3].

3. Process for Simultaneous SO_2- and NO_x- Removal

The process runs under conditions usually prevailing downstream of the air preheater and the electrostatic precipitation. The temperatures may vary between 100 and 170 °C.
The process for simultaneous SO_2- and NO_x-removal from flue gases of coal or oil fired boilers is applied industrially. Process modifications are developed for an improved clean-up of off-gases from waste incineration plants.

3.1 Process for Simultaneous SO_2- and NO_x-Removal

At first the process for the desulphurization of flue gases from utility boilers and power plants was developed. One-stage adsorbers are used for SO_2-adsorption on active coke.
This process was improved by the integration of a second stage where active coke is used as catalyst for NO_x reduction. The latter process for simultaneous SO_2- an NO_x-removal is described in the following.
The adsorber is installed downstream of the electrostatic precipitator and the blower. A two-stage moving bed adsorber with a cross-flow of the flue gases is used.

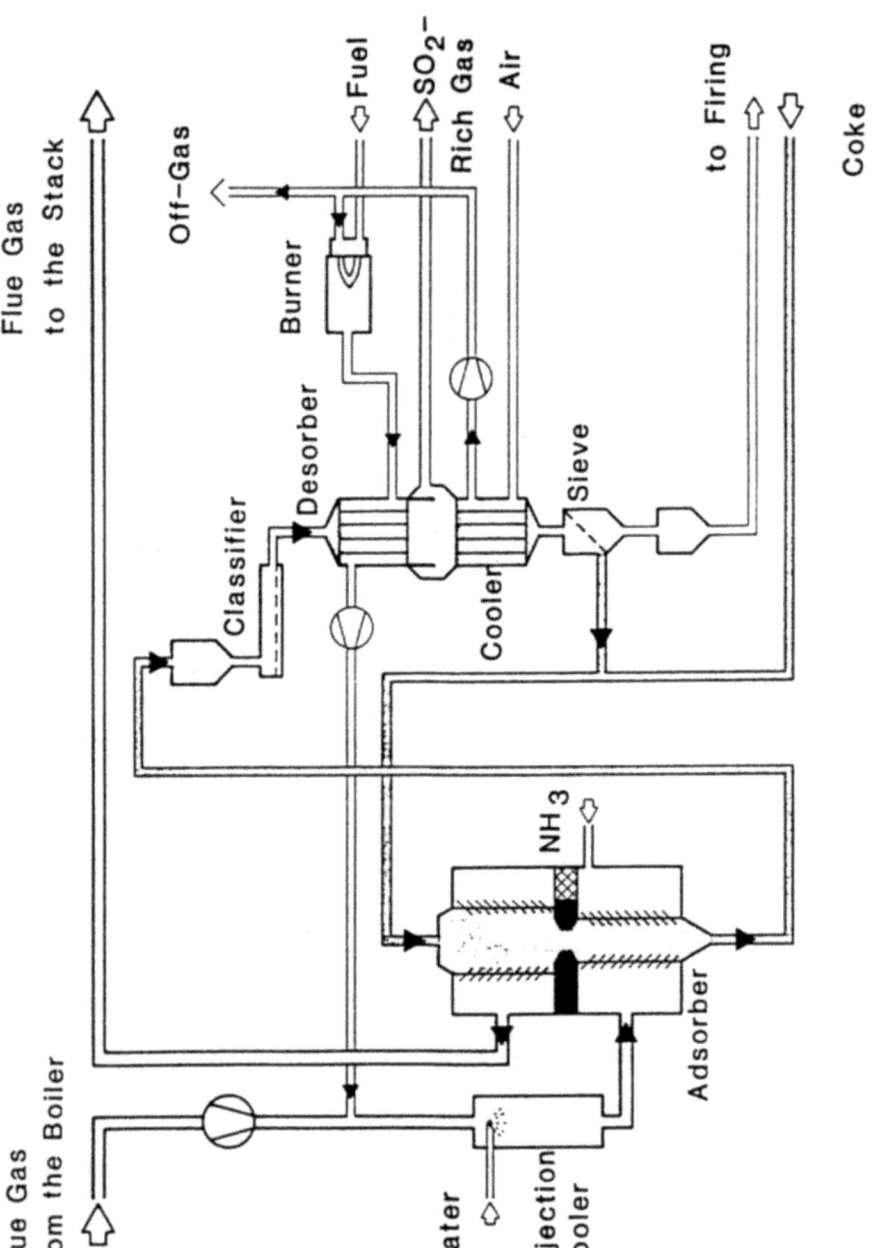

Fig.1 Simplified flow sheet of the BF/UHDE/MITSUI-Process for SO_2/NO_x-removal: Adsorption and regeneration section

Active coke moves from the top to the bottom of the two stages which can be arranged one upon another. A simplified flow-sheet of the process is shown in Fig.1. The first (lower) bed serves for the removal of a major part of the sulphur dioxide (according to eq.(1)) and nitrogen dioxide. No ammonia will be added upstream of this bed since the adsorption capacity of active coke for SO_2 resp. sulphuric acid is sufficient.

Prior to the second (upper) bed ammonia will be injected for the reduction of nitric oxide in the catalyst bed (according to eq.(2)). Furthermore the residual sulphur dioxide will be removed (according to eq.(1) and (3)).

As the adsorbent is only slightly charged with sulphur compounds in the upper moving bed, the adsorbent leaving this top bed can also be used in the lower bed for SO_2 separation. In this way, the adsorbent can be charged to a greater degree than in the single-unit method.

Besides SO_2 and NO_x, other gaseous components of the flue gases are removed by active coke: hydrogen chloride, hydrogen fluoride, mercury, arsenic and selenium compounds.

The charged active coke is thermally regenerated for reuse. A tube desorber, a simplified diagram of which is shown in Fig.1, may be used for this. The adsorbent is heated up when flowing through the upper part of the tubes. When it reaches 400 to 450 °C it outgases in a middle section. The SO_2 enriched gas released contains 25 to 30 % by volume of SO_2, which is cleaned in a special unit and processed further to S, H_2SO_4 or liquid SO_2. Before returning to the moving beds, the adsorbent is cooled down to approximately 100 °C in a tube cooler. Air is used as cooling agent which is fed into the burner which produces the combustion gases for heating-up the active coke in the upper section of the desorber. The regenerated active coke is transported back into the adsorber. Adsorption and regeneration need not be carried out at the same place. This system can not only be installed in new power stations, but can also easily be fitted subsequently to existing power stations as it can be integrated into the flue gas stream between the electrostatic precipitator and the stack.

As a rule over 95 % of SO_2 is removed (as a consequence of the fact that ammonia reacts faster with sulphur dioxide than nitrogen oxides in the second stage). The reduction rate of the nitrogen oxides depends on the design of the second stage. In the Fed. Rep. of Germany the nitrogen oxides must be reduced to concentrations below 200 mg NO_2/m^3. This is possible even for the flue gases of boilers with fluid slag removal. Furtheron, the chlorine, fluorine, mercury, arsenic and seleninium compounds and parts of the dust remaining after electric precipitation are reduced.

The SO_2 rich gas contains the desorbed gaseous components. Besides SO_2, water vapour and carbon dioxide minor concentrations of the halogens, mercury, arsenic, selenic and ammonia are observed. The latter are removed in a separate stage. The gas is led through two scrubbers to remove As_2O_3, Se, Hg and ammonia by reaction with sulphuric acid. The halogens are dissolved in the water condensate produced in the subsequent gas cooler. The condensate is neutralized by sodium hydroxide an removed to the flue gas quencher or it is vaporized e.g. cristallized. Sulphuric acid aerosols are removed by electric precipitators. The last unit is a safety filter containing activated carbon for the removal of possible traces of mercury. The solutions of the two scrubbers are circulated. Solid products are removed from the solution by a filter press. The filter cake has to be disposed in a special disposal area. By this method a complete removal of these substances is

possible without pollution of rivers etc. The latter may occur by waste waters of the wet flue gas desulphurization processes.
Processing of SO_2 rich gas can lead to sulphur, sulphuric acid or liquid SO_2. In every case, the above described gas treatment is necessary to remove hydrogen chloride, hydrogen fluoride, dust and heavy metals if necessary. For sulphur production, parts of the SO_2 are reduced to S and H_2S in a reduction burner. H_2S and SO_2 are converted to sulphur in a subsequent Claus unit. From condensers, liquid sulphur is drawn off. Sulphurous compounds in the off-gases are oxidized in a final combustion unit and fed to the flue gas desulphurization stage [4].
Concentrated sulphuric acid is produced by catalytic oxidation after addition of air. The tail gas is fed to the flue gas desulphurization stage. Processing of SO_2 rich gas to liquid SO_2 contains coolers and dryers and a CO_2 stripping column.
The lay-out data for the three plants already operated are shown in <u>Table 1</u>. In every case, SO_2 is removed almost completely. The design of the NO_x-stage is adapted to the necessities of NO_x emission control.
The process is a true process for simultaneous SO_2- and NO_x-removal. It produces no waste water and processes sulphur to valuable end products. Its introduction to the market was hindered by the sequence of the emission regulation steps. In the Federal Republic of Germany, at first only SO_2 from half of the flue gases had to be removed. So the wet processes established which were much cheaper as no reheating of the flue gases was necessary under these conditions. The problems of waste water had not been recognized at that time. The costs for NO_x-removal were not included in cost calculation. Nowadays, all the additional measures are retrofitted to the flue gas cleaning plants. All these measures are not necessary for the BF/UHDE/MITSUI-Process as they are included in the original processes. As a result of the German experience, it must be recommended to other European countries to introduce regulations for flue gas cleaning not step by step but in one step so that the best over-all solution of every power plant can be found.

3.2 Improved Clean-Up for the Off-Gases from Waste Incinerators

The off-gases of waste incineration plants do not only include SO_2 and NO_x but also HCl, HF, heavy metals and dioxines and furanes. Until now, waste incineration plants are equiped only with electrostatic precipitators and, in some cases, with wet or dry scrubbers for HCl- and SO_2-removal. Even these scrubbers are insufficient to meet the regulations which are introduced or expected in Austria, The Netherlands and the Federal Republic of Germany [5].
The following dayly and half-hour average emissions are expected in the Federal Republic of Germany:

		dayly average	half-hour average
Dust	(mg/m^3)	10	60
HCl	(mg/m^3)	10	60
HF	(mg/m^3)	1	4
SO_2	(mg/m^3)	50	200
NO_2	(mg/m^3)	100	400
Hg,Cd,Tl	(mg/m^3)	0.2 (Average over sampling period)	
Dioxine and Furanes	(ng/m^3)	0.1 (Toxicity equivalent)	

Table 1

BF/UHDE/MITSUI-Process for SO_2- and NO_x- removal from flue gases: Plants in operation

Supplier	Uhde GmbH		Mitsui Mining Comp	Uhde GmbH	
Plant	Kraftwerk Arzberg		Aichi Refinery	Hoechst AG	
	Lignite Fired Boiler		FCC Off-Gases	Trivalent Boiler	
	Block5	Block 7		Hard Coal	Gas
Flue Gas Volume (m^3/h (STP))	451 000	658 000	236 000	323 000	278 000
Temperature (°C)	150	150	190	140	120
Inlet:					
SO_2 (mg/m^3, dry)	3 800	3 600	760	2 500	2
SO_3 (mg/m^3, dry)	125	125	110	210	-
NO_2 (mg/m^3, dry)	750	750	205	900	450
dust (mg/m^3)					
Removal Efficiencies:					
SO_x (%)	>95	>95	>90	>90	-
NO_x (%)	>73	>73	>60	>78	>75
Residual Contents:					
Dust (mg/m^3)	50	50	20	50	-
NH_3 (mg/m^3, dry)	<35	<35	<35	<35	<35
Start-up	7/1987	9/1987	5/1987	9/1989	

Active coke plants shall be used for the removal of heavy metals, dioxines and furanes, and NO_x downstream of scrubbers. An improved removal of SO_2, HCl HF and dust is also obtained by this technology. This process is also operated as a two-stage process, as shown on Fig.2.

The lower stage serves for the removal of gaseous heavy metals (especially Hg) and of dioxines and furanes. It is really surprising that the organic components can be removed to values under 0.1 ng/m^3. In this stage, residual SO_2, HCl and HF, which are not removed by the scrubber, are adsorbed rather completely. Ammonia is added to the gas stream before it enters the upper stage where NO_x-removal is carried out. The loaded coke is withdrawn from the lower stage and transported to a desorber which is operated in an oxygen-free atmosphere. The desorber size is very small because the active coke volume to be regenerated is small due to the removal of contaminants prevailing at low concentrations.

The dioxines and furanes are decomposed at temperatures above 400 °C in this oxygen-free gas. The mercury-compounds are desorbed as well as SO_2, HCl and HF. It is possible to remove the heavy metals before the gas is recycled to the off-gas stream of the waste incineration plant upstream of the electrostatic precipitator.

A variant of this process uses calcined lignite in a first stage moving bed and active coke in a second stage fixed bed. A plant, following this concept, is under construction at one site in The Netherlands. The design has to meet the above mentioned expected regulations for the Federal Republic of Germany but lower emissions for NO_x (70 mg NO_2/m^3).

4. Conclusions

Cost calculations for the BF/UHDE/MITSUI-Process for simultaneous SO_2/NO_x-removal show the following result (based on costs in 1987) [6]: The investment costs are about 330 to 420 DM/kW_{el}. The operating costs (5000 hours/year) are about 0.020 DM/kWh_{el}. Only about 27 % of the operating costs are necessary for the reduction of NO_x. Based on market experience, the operating costs for the simultaneous process are comparable of even lower than the costs for compceible processes, e.g. wet lime-based SO_2-srubbing with gipsum production combined with a SCR-process for NO_x-reduction.

The BF/UHDE/MITSUI-Process for simultaneous SO_2- and NO_x- removal is one of the most advanced processes for flue gas cleaning, removing not only SO_2 and NO_x but also HCl, HF, selenium and arsenic compounds, mercury and dust from the flue gases. The main advantages of the process are:
- High separation efficiencies for SO_2 and NO_x
- Usable end products: sulphur, H_2SO_4 or SO_2 (liquid), hence no dumping problems
- No waste water, hence, problems are not transfered from the flue gas into the waste water.

These advantages can be achieved with operating costs which are comparable or even lower than those for competitive processes.

165

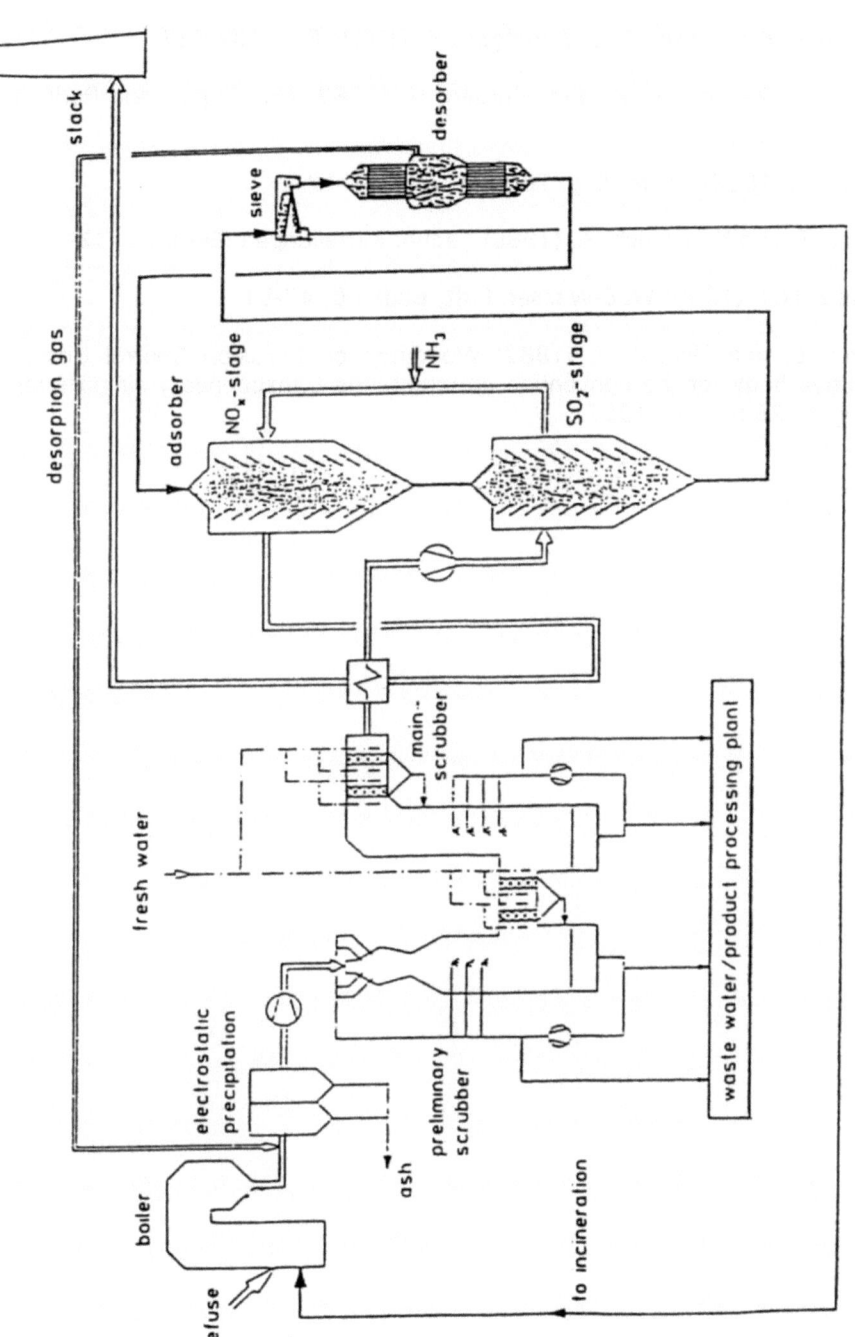

Fig. 2 Combination of wet scrubber and two-stage active coke process for clean-up of the off-gases from waste incineration plants

References

1) Knoblauch, K.; Richter, E. and Jüntgen, H. (1981) Fuel 832-837

2) Richter, E.; Knoblauch, K. and Jüntgen, H. (1987) Gas Separ. and Purif. 1, 35-43

3) Richter. E. (1990) Chem. Eng. Technol. 2, 101-112

4) Richter, E. and Knoblauch, K. (1985) Techn. Mitteilungen (Essen) 1, 13-15

5) Schmitz, H.J. (1990) WLB-Wasser, Luft, Boden 6, 47-51

6) Richter, E. and Henkel, J. (1987) Workshop on Emission Control Costs - Executive body for the convention on long-range transboundary air pollution, Esslingen, 28.9.-1.10.1987

WALTHER PROCESS

W. SCHULTE
Krupp Koppers GmbH
POB 10 22 51 Essen
Federal Republic of Germany

1. Introduction

   The Walther process for the desulphurization of flue gases was developed with the aim of establishing a desulphurization method with the following features:

   - avoidance of waste water and other waste products
   - production of the marketable nitrogen fertilizer ammonium sulphate from the sulphur dioxide contained in the flue gases
   - reheating the clean gases without external heating energy

   These important characteristics of the Walther process were first demonstrated in 1978 in a pilot plant for a throughput of 20,000 Nm^3/h.

   This pilot plant was operated first downstream of an oilfired boiler and thereafter downstream of a coalfired boiler.

   A non-objection certificate for the use of ammonium sulphate resulting from flue gas desulphurization as a fertilizer was issued in 1980 by the University of Hohenheim in the Federal Republic of Germany.

   The satisfactory results of the pilot plant operation led to the implementation of a large scale plant with a flue gas throughput of 750,000 Nm^3/h.

   This plant was put into operation in 1983. The separation of sulphur dioxide was completely satisfactory and all other process guarantees regarding the clean gas properties were met.

   The ammonium sulphate produced in the form of pellets was successfully marketed and readily accepted by the market. Due to its superior handling properties the demand increased over and above the supply potential for this type of fertilizer.

   For reducing the aerosol content of the clean gas, which led to an aerosol plume at the stack, a special filter system was developed and installed.

   This aerosol filter has been in operation since 1986 and the aerosol separation system has proven to be completely satisfactory under varying operating conditions.

   The operational results of the large scale plant led to a further contract for a flue gas desulphurization plant in 1986. This FGD plant is combined with a NO_x removal system to be installed downstream of the FGD plant.

This plant, which was installed at Karlsruhe, is now in operation and has shown satisfactory results.

A demonstration facility to desulphurize flue gases containing up to 16,000 mg/Nm^3 SO_2 is currently being constructed in Italy.
This plant will be put into operation in November 1990.

2. Special Process Features and Advantages

The Walther process for desulphurization of flue gas using ammoniacal scrubbing has the following characteristics and special advantages:

Flue gas desulphurization according to the Walther process reliably achieves the required SO_2 limits in the off-gases independent of the inlet concentration.

Scrubbing with ammonia water produces ammonium sulphate, a marketable fertilizer which has become well established meanwhile in the fertilizer market. The Walther process does not create waste disposal problems or market difficulties with the by-products such as occur with other processes. Krupp Koppers can assist in arranging long-term contracts which assure the supply of ammonia and the acceptance of the ammonium sulphate.

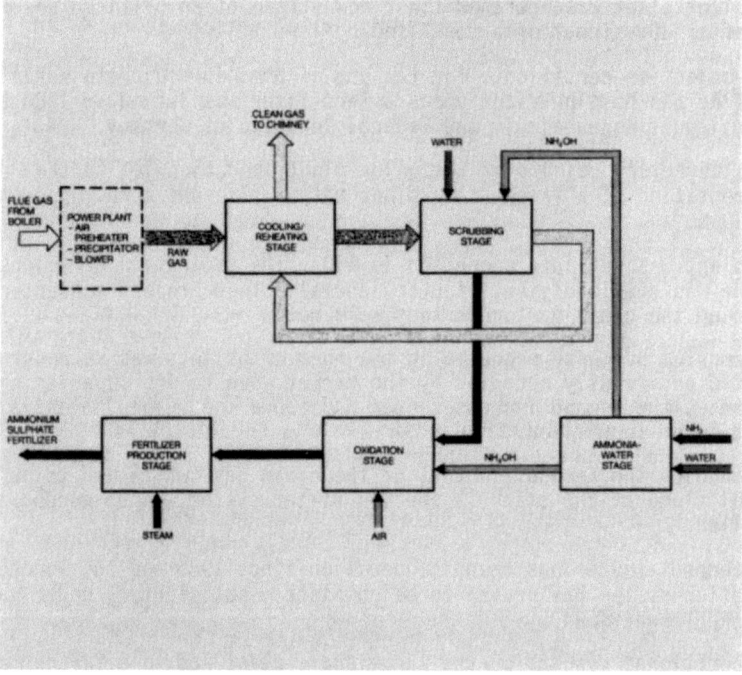

Fig. 1   WALTHER FGD - PROCESS

The Walther flue gas desulphurization process produces no waste water. All resulting flushing water is collected and recycled into the system as process water.

The reheating of the cold clean gases to the required stack inlet temperature is easily effected by means of heat exchange with the untreated flue gas, i.e. without external heating.

3. **Process Description**

In the WALTHER flue gas desulphurization process the flue gases are scrubbed with ammoniacal solution in order to extensively remove SO_2, SO_3, HCl and HF. Absorbed in the solution, these pollutants react according to the following equations:

$$SO_2 + (NH_4)_2SO_3 + H_2O \longrightarrow 2NH_4HSO_3 \qquad \text{ammonium hydrogen sulphite} \quad (1)$$

$$SO_3 + 2(NH_4)_2SO_3 + H_2O \longrightarrow 2NH_4HSO_3 + (NH_4)_2SO_4 \qquad \text{ammonium sulphate} \quad (2)$$

$$HCl + (NH_4)_2SO_3 \longrightarrow NH_4HSO_3 + NH_4Cl \qquad \text{ammonium chloride} \quad (3)$$

$$HF + (NH_4)_2SO_3 \longrightarrow NH_4HSO_3 + NH_4F \qquad \text{ammonium fluoride} \quad (4)$$

The ammonia consumed in this way is passed to the scrubbing cycle in the form of ammonia liquor.

$$NH_4HSO_3 + NH_4OH \longrightarrow (NH_4)_2SO_3 + H_2O \qquad \text{ammonium sulphite} \quad (5)$$

In addition, the salts react with the oxygen contained in the flue gas as follows:

$$2(NH_4)_2SO_3 + O_2 \longrightarrow 2(NH_4)_2SO_4 \qquad \text{ammonium sulphate} \quad (6)$$

In the oxidator the discharged scrubbing solution reacts as per equation (6) with atmospheric oxygen. Ammonium hydrogen sulphite is likewise converted to ammonium sulphate by adding ammonia liquor and by means of oxidation with atmospheric oxygen:

$$2NH_4HSO_3 + O_2 + 2NH_4OH \longrightarrow 2(NH_4)_2SO_4 + 2H_2O \qquad \text{ammonium sulphate} \quad (7)$$

The ratio of ammonium sulphite to ammonium hydrogensulphite in the scrubbing solution determines the degree of absorption, whereby the pH-value is an indirect indicator of this proportional relationship. In the case of lower pH-values (a lower proportion of ammonium sulphite), the absorption capability is reduced and rises in the case of higher pH-values (higher proportion of ammonium sulphite).

Under specific temperature and concentration conditions, ammonium salts - the so-called aerosols - may also form in the gas phase.
The formation and separation of aerosols will be described later in this paper.

4. Scrubbing System and Oxidation Unit

After cooling in a heat exchanger the dedusted flue gas is fed to the head of scrubber 1. Scrubber 1 has no internal fittings and works in co-current flow. The scrubbing solution, in which the reaction products are dissolved, is withdrawn from the bottom and sprayed at various levels in the top of the scrubber into the flue gas stream. To this scrubbing solution ammonia water is added as absorbent.

The pH-controlled scrubbing solution is circulated, whereby a constant salt concentration is maintained.

The flue gas charged with scrubbing solution leaves scrubber 1 at the bottom and is fed via a droplet separator to scrubber 2.

In scrubber 2 the flue gas is washed once more. The scrubbing solution is ammonia water with a lower concentration of reaction products.

Downstream of scrubber 2 the flue gas still contains fine dust, very fine droplets and aerosols. Depending on particle concentration and their size distribution the particles are separated in droplet and aerosols separation systems and returned to the scrubbing system in the form of a solution.

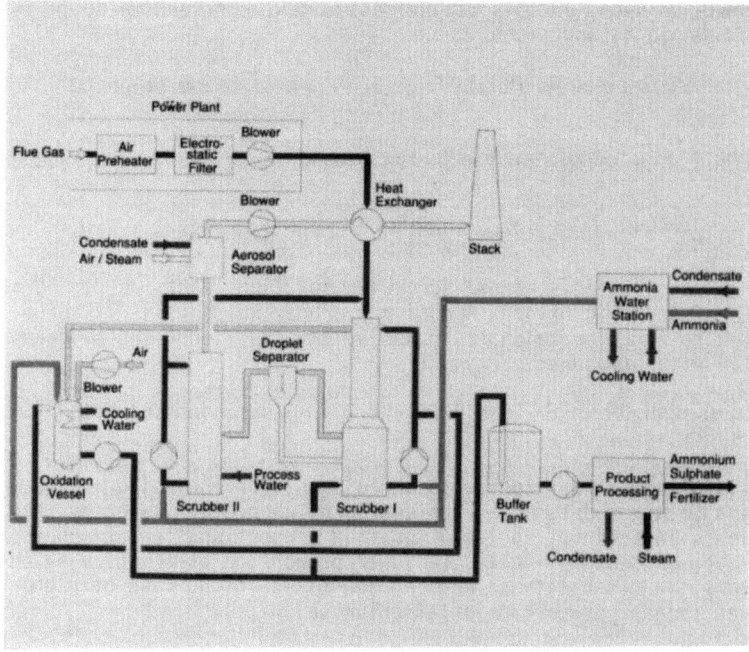

Fig. 2 SCRUBBING AND OXIDATION STAGE

The desulphurized flue gas is conveyed via a blower to a heat exchanger, heated up by hot non-desulphurized flue gas and discharged at a temperature of 75 °C via a stack into the atmosphere.

A part of the scrubbing solution, containing reaction products, is withdrawn from scrubber 1 and thoroughly mixed with air in an oxidation vessel. At the same time the required pH-value is regulated and kept constant by adding ammonia water. In this way ammonium hydrogen sulfite is converted to ammonium sulphite. The ammonium sulphite is oxidized by atmospheric oxygen to ammonium sulphate.

Gases leaving the oxidation vessel are recycled into scrubber 1.

The aqueous solution leaving the oxidizing tank and containing ammonium sulphate and small quantities of ammonium chloride and ammonium fluoride is stored in a buffer tank.

5. Ammonium Sulphate Processing

Concentration

The aqueous ammonium sulphate solution flows from the buffer tank of the oxidizing section to the concentration section. The solution is preheated by vapour condensate in the preheater before it is fed into the evaporation circuit. The recirculation heater is heated by vapours, the pressure of which has been elevated by a vapour compressor. The vapours separated from the solution in the evaporator are scrubbed and compressed in the vapours compressor before being used as heating medium. For start-up and partial load conditions preheating is possible.

The resulting crystal suspension is pumped from the evaporator to a hydrocyclone. The suspension with the required concentration for the granulation process flows as under-flow into the granulator feed tank.

Granulation

Before the crystal suspension is fed to the granulator, additives can be dosed to improve the fertilizer properties, if required. The suspension containing very fine crystals is sprayed onto a moving bed of granules in the granulating drum. The granulator is a horizontal drum fitted with lifters of a special anticlogging design and an integrated fluidized bed which is supplied with pre-heated air.

The lifters elevate the product contained in the granulating drum and let it flow onto the fluidized bed surface. The product is dried on the fluidized bed by the pre-heated air - whereby the heat of crystallization is also utilized - and falls back to lower part of the granulating drum.

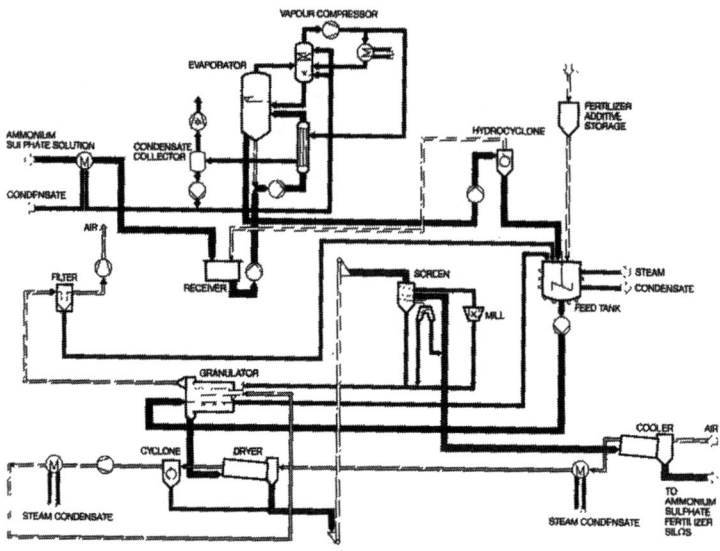

Fig. 3   FERTILIZER PRODUCTION STAGE

There the product is sprayed with the suspension and the granules are increased in diameter due to the successive layer formation. The granules follow this cycle a number of times before they leave the granulator. The moisture of the granules is very low because of the intensive evaporation occurring in the fluidized bed. Final drying according to the requirements takes place in the downstream product dryer.

The granules are then fractionated on the sizer. The oversize product is separated on the sizer and after grinding in the rolling crusher recycled to the elevator. The separated fines are sent directly to the granulator as seed granules.

The on-size granules are cooled in the product cooler and then transported pneumatically into the fertilizer silos. The cooling air for the product cooler is preheated and used in the product dryer. The finished product leaves the fertilizer silos via the bulk loading equipment.

All off-gases of this plant section are dedusted by filters. The dust returns either into the respective silos or into the reslurrying vessel.

6. Ammonia Water Station

Liquid ammonia is supplied in rail tank cars.

During unloading the tank car is connected with the gas pipe system as well as with the liquid pipe system of the ammonia unloading station. Ammonia is withdrawn from the connection for liquid ammonia of the tank car and pumped into the storage tanks.

6.1 Ammonia storage

Ammonia is stored in horizontal storage tanks which are equipped with a water spraying system. The spraying system is necessary in case of ammonia leakage and fire as well as cooling in case of solar irradiation.

As it is possible that the spraying water contains dissolved ammonia, it is collected in a pit and fed to the circuit of scrubber 1.

6.2 Preparation of ammonia water

Liquid ammonia is conveyed by pumps from the storage tanks to the ammonia water preparation unit.

Water (vapour condensate) and liquid ammonia are mixed at a pressure of approx. 1 bar above the vapour pressure of ammonia. In this way the formation of gas bubbles is avoided.

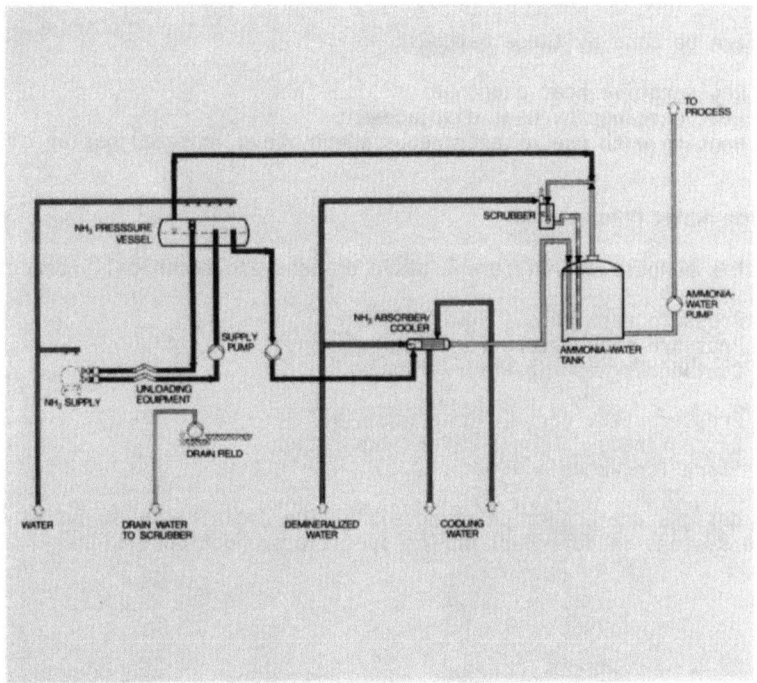

Fig. 4   AMMONIA-WATER PREPARATION STAGE

Liquid ammonia and water are mixed in a mixer/cooler system by means of a ratio control device.

The ammonia water with an ammonia content of 25 wt-% is stored in a vertical flat bottom tank.

7. Process Design Variations

Reheating of Clean Gas (Desulphurized Flue Gas)

Depending on fuel and type of boiler process the temperatures downstream the heat exchangers of the power plant are 130 ° up to 170 °C.

The temperature at the stack outlet stipulated in the "Großfeuerungsanlagenverordnung" (GFAV) (power plant regulations) is 72 °C.

As the Walther desulphurization process uses the principle of wet scrubbing and operates in accordance with adiabatic principles, it is necessary to reheat the purified gases.

This can be done by three methods:

a) Regenerative heat exchange
b) Heat exchange by heat displacement
c) Heating with use of extraneous energy (e.g. natural gas or oil)

7.1 Ammonia Water Preparation

For this purpose two different basic designs are technically practicable:

a) Pressure tank cars (liquid ammonia)
   Pressure tank (store for liquid ammonia)
   Continuous ammonia water preparation

b) Pressure tank cars (liquid ammonia)
   Discontinuous ammonia water preparation
   Store for ammonia water

Both designs are economically possible. The decision in favour of one of these systems is dependent on the specific project conditions.

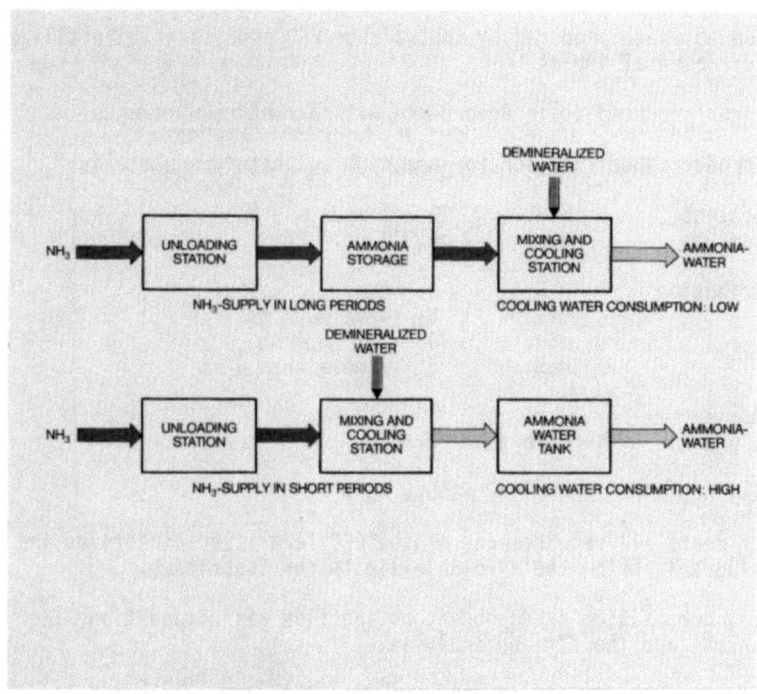

Fig. 5   AMMONIA-WATER PREPARATION STAGE

8. Process Application Possibilities

In principle it is possible to use the Walther-desulphurization process in all cases where SO_2-containing gases occur independent of the origin or kind of formation, as for example different methods of firing (slag tap furnace, dry bottom furnace) usually employed in the field of power plants.
The suitability of application of the Wather-process downstream of sintering belt plants, refuse incineration plants, brown coal-fired boilers or smaller power plants has to be studied in each case.

9. Typical Consumption Figures

Basis: Hard coal fired boilers
  Power rating: 100 MW
  Flue gas flow: 300,000 m³/h
  SO_2-content (inlet)   2,200 mg/m³
  SO_2-content (outlet)   200 mg/m³
  Ammonium sulphate output: 1,361 kg/h

Consumption per hour
  Liquid ammonia   351 kg
  Process water   5 m³
  Cooling water   80 m³
  Electric power   2,000 kWh
  Steam (medium pressure)   1.5 t
  Demineralized water   3 m³

10. Ammonium Sulphate Fertilizer

The ammonium sulphate produced by the Walther FGD process is a fertilizer with no environmental impact.

The fertilizer produced is in accordance with market requirements.

A typical product specification for ammonium sulphate granulate is:

Nitrogen content	min.	20	% wt.
Water content	max.	0.3	% wt.

Grain distribution

max.	5	% below 2 mm
min.	90	% 2 to 4 mm
max.	5	% more than 4 mm

Grain hardness
(of a 3 mm grain)   more than 2.5 kg

Bulk density   800 kg/m^3

The product meets all requirements of the EEC fertilizer directives and falls considerably below the allowable limits for impurities.

The chemical composition is dependent on the flue gas composition, the fly ash content and the fly ash analysis.

The impurities of the fertilizer produced are very low in comparison with other fertilizers produced from natural materials available on the market.

Furthermore, by using additives, the nutrient yield can be increased and an undesirable quick transition of nitrate nitrogen resulting from the ammonia nitrogen from the soil into the groundwater can be considerably reduced.

11. Economic Aspects

Depending on the specific project conditions the desulphurization costs will vary to a certain extent and have to be calculated in each individual case.

However, the desulphurization costs are influenced by certain process features.

The decisive factors which favor the Walther process are:

- It can be assumed that the reacting media ammonia will be provided free of charge in exchange for the produced ammonium sulphate fertilizer at plant site.

    This is confirmed by the existence of long term contracts between power plant companies and fertilizer producers in Germany.

- When using a limestone scrubbing process additional cost items have to be considered

    o   Limestone cost
    o   Transportation cost for limestone
    o   Transportation cost for gypsum
    o   Water treatment cost
    o   Sludge disposal cost

The credit for the gypsum produced, based on German conditions, is only a small fraction of these costs.

For these reasons the desulphurization costs for the Walther process are substantially lower than for a limestone scrubbing process, which makes the Walther process favourable from the economic point of view.

12. References

The results obtained at Großkraftwerk Mannheim AG led to an order for the turnkey construction of a flue gas desulphurization plant for a power boiler of Stadtwerke Karlsruhe.

The order was awarded in December 1986 and the plant has been put into operation mid-1988.

A further order is currently in the engineering phase: Krupp Koppers is building a demonstration plant for the Italian utility group ENEL in Sulcis, Sardinia, for the desulphurization of power plant flue gases with SO_2-concentrations up to 16,000 mg/m^3.

This plant will be put into operation in November 1990.

13. Formation of Aerosols

Aerosols can be formed in ammoniacal scrubbing processes under certain conditions. We define aerosols as droplet or solids particles, with a diameter of less than 1 micron.

Their formation process can be explained as follows:

$$NH_3 + SO_2 + H_2O \longrightarrow NH_4HSO_3 \qquad (1)$$

$$2\,NH_3 + SO_2 + 2\,H_2O \longrightarrow (NH_4)_2SO_3 \times H_2O \qquad (2)$$

The course of the reaction is calculated with the help of the equilibrium constant "K", whereby the constant is a function of the temperature.

$$K_1 = f(T) \qquad (3)$$

$$K_2 = f(T) \tag{4}$$

Reactions (1) and/or (2) may occur if the values π_1 / π_2 are greater than the values K_1 / K_2.

Values π_1 / π_2 are determined according to the following equations:

$$\pi_1 = P_{NH3} \times P_{SO2} \times pH2O \tag{5}$$

$$\pi_2 = P_{NH3}^2 \times P_{SO2}^2 \times pH2O^2 \tag{6}$$

The reactions of aerosol formation from SO_3, HCl and HF proceed as follows:

$$SO_3 + H_2O \longrightarrow H_2SO_4 \tag{7}$$

$$H_2SO_4 + 2NH_3 \longrightarrow (NH_4)_2SO_4 \tag{8}$$

$$H_2SO_4 + NH_3 \longrightarrow NH_4HSO_4 \tag{9}$$

$$HCL + NH_3 \longrightarrow NH_4Cl \tag{10}$$

$$HF + NH_3 \longrightarrow NH_4F \tag{11}$$

The reactions take place in the gas phase.

For this reason it is not always necessary to employ an aerosol separating device. For example, no aerosol separation was required in the flue gas desulphurization facilities of the Stadtwerke Karlsruhe (municipal utilities). In this plant the criteria decisive for aerosol formation are so favourable that when observing specific operating conditions virtually no aerosol fumes occur here at all.

14. Aerosol Separation

The behaviour of aerosol particles is determined by their size. The smaller the particle, the more its movement resembles the pattern of "Braunian motion". As diameter decreases, such particles are more and more subject to the laws of gas diffusion. The best method of separating them is therefore if they remain as long as possible in a dense packing with the shortest possible path lengths.

The motion of larger particles (approx. 0.1 to 1.0 μm) is only slightly affected during the impulse exchange with gas molecules, more and more they tend to follow the "filaments of flow".

When considering aerosol separation the aerosols are present with a particle size of 0.1 - 1 micron.

In the WALTHER process the aerosols formed in the gas phase are highly water-soluble solids, which also exhibit hygroscopic properties. They are therefore capable of condensing water on their surface from a steam-laden gas atmosphere. Should this occur, droplets of increasing diameter result in which the original solids particle is dissolved.

As described above, in a saturated atmosphere hygroscopic aerosols envelop themselves in water. According to the mechanism of cloud formation in the earth's atmosphere the very fine aerosols can be increased by droplet formation, if enough liquid potential is provided to maintain saturation by feeding finely dispersed water to the gas phase. Ideally this water would have to be salt-free or at least have a lower salt content than the formed aerosol droplet, so that the propulsion is maintained for condensation on the saliferous droplet.

Systematic investigations of aerosol growth rate have indicated that under corresponding conditions, the time required for the aerosol to increase from 0.1 to 1.0 μm, i.e. a thousand times original volume, is less than 1 second.

Various systems are currently available for aerosol separation:

- Mist eliminator (e.g. Brink type)
- Wet-type electrostatic precipitator
- Multi-stage separation, comprising:

  Large droplet separator Condensation stage Agglomeration stage fine droplet separator

15. Combining the WALTHER Flue Gas Desulphurization Process with DENOX Facilities

Since ammonia is used as absorbent in the WALTHER desulphurization process, it would be worth considering to what extent ammonia can also be employed for nitrogen removal. The following processes are available:

- SCS
- SNCR
- WALTHER simultaneous separation

Since in the SNCR process it is necessary to intervene in boiler operation, we did not feel that we had the necessary experience as we are not boilermakers.

The WALTHER simultaneous process works on the following principle: after the desulphurization stage the NO is oxidized to NO_2 and N_2O_5 by injecting ozone into the gas phase which is then absorbed in the ammoniacal solution. It emerged, however, that due to the high energy requirement for ozone production, the cost of utilities was too high in this process

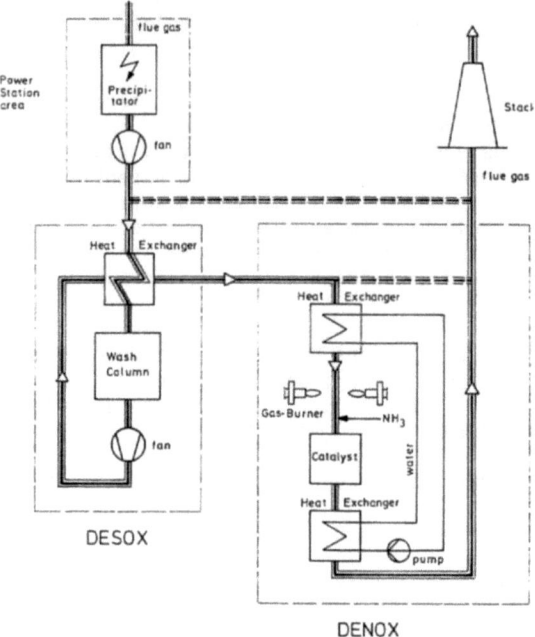

Fig. 7   COMBINED DESOX-DENOX
STADTWERKE KARLSRUHE

as compared to the SCR technology. In this case price development for
catalysts was to be taken into account, particularly against the background of constantly falling prices. Another factor to be considered was
that catalysts are remaining in service for much longer periods than was
previously the case.

For this reason development work is being performed at Krupp Koppers to
optimize the simultaneous process. No results can be published yet as
this work has not been completed.

Under consideration of these facts in the case of the Stadtwerke Karlsruhe project an SCR DENOX unit was foreseen in addition to and installed
downstream of the WALTHER facilities.

The gases from which SO_2 has been largely removed leave the desulphurization unit at a temperature of approx. 90 °C and are passed to a heat
transfer medium. Additional heating of the gases by approx. 20 °C is
effected by means of natural gas. The required amount of ammonia is then
passed to the flue gas via a distribution system. The reduction to form
nitrogen oxides takes place in the downstream honeycomb catalyst at a
temperature of 320 °C. Using the heat displacement system the clean gases
are cooled and routed to the stack at a temperature of approx. 130 °C.
The combination of a WALTHER desulphurization unit and an SCR system is
possible not only in this configuration, but just as easily in the
reverse configuration, i.e. the SCR system can be installed upstream of
the flue gas desulphurization unit, whereby both the so-called high dust
and low dust configurations can be considered.

The conditions specific to the project have to be taken into account when
selecting a particular configuration, whereby decisive factors are above
all cost of utilities and space available. We consider the combination of
the WALTHER desulphurization process with the SCR unit to be particularly
favourable, since the power plant operator requires only one chemical,
i.e. ammonia, to solve both problems, namely removal of SO_x and removal
of NO_x.

EBDS-Process

H.-R. Paur

Kernforschungszentrum Karlsruhe GmbH

Laboratorium für Aerosolphysik und Filtertechnik I

Postfach 3640

D-7500 Karlsruhe 1

1. Introduction

SO_2 and NO_x emissions from power plants are partly responsible for acidic precipitations, which lead to forest decay, crop damage and building corrosion. The amounts of these emissions for industrialized countries are in the order of 3 Mio tons/year for West-Germany and 23 Mio. tons/year for the U.S.A. In order to decrease the substantial damages from acidic precipitation, many countries are presently imposing stringent emission standards for fossil fuelled power plants. Conventional technologies which are capable of meeting these standards such as wet scrubbers for SO_2 absorption and SCR for NO_x reduction are meanwhile available. Whereas being quite suitable for large power plants, these technologies become less cost effective when scaled down to smaller units. This is especially the case for wet scrubbing systems, which are highly complex and require sophisticated waste water and product treatment.

In the EBDS process NO_x and SO_2 are oxidized into HNO_3 and H_2SO_4 by radicals (OH, O_2H), which are generated due to the interaction of accelerated electrons with the main components of the flue gas such as N_2, O_2, CO_2 and H_2O [1-3]. The gas phase chemistry of the process therefore partly resembles the reaction sequences, which have been proposed for the formation of the photochemical smog. The accelerated electrons are supplied by commercial high power accelerators which have an output of up to 600 kW. Since H_2SO_4 has a very low vapour pressure at the typical operating conditions of the EBDS process (70° C; $[H_2O]$ = 10-15 Vol.%), it nucleates readily to form a submicron aerosol. These aerosol droplets are then neutralized by the injection of stoichiometric amounts of ammonia. The aerosol may be filtered from the cleaned flue gas using fabric filters or electrostatic precipitators (ESP). The product is a mixture of ammonium sulfate and nitrate and can be used for the production of fertilizer.

The EBDS process has several advantages: It is a dry process, thus requiring no waste water treatment. The reaction product is a high quality fertilizer. Due to the quick response of the electron accelerator, the process is very flexible and can rather easily follow load fluctuations, which occur frequently in smaller power plants. The process removes NO_x and SO_2 simultaneously in one system and has low space requirements. Pilot scale test plants have been operated during the last years in Germany, Japan and the U.S.A. The sizes of these plants range from 1000 m3/h up to 30.000 m3/h. From the test results of these installations the knowledge regarding chemistry, physics and engineering of the EBDS process has advanced considerably [4-9].

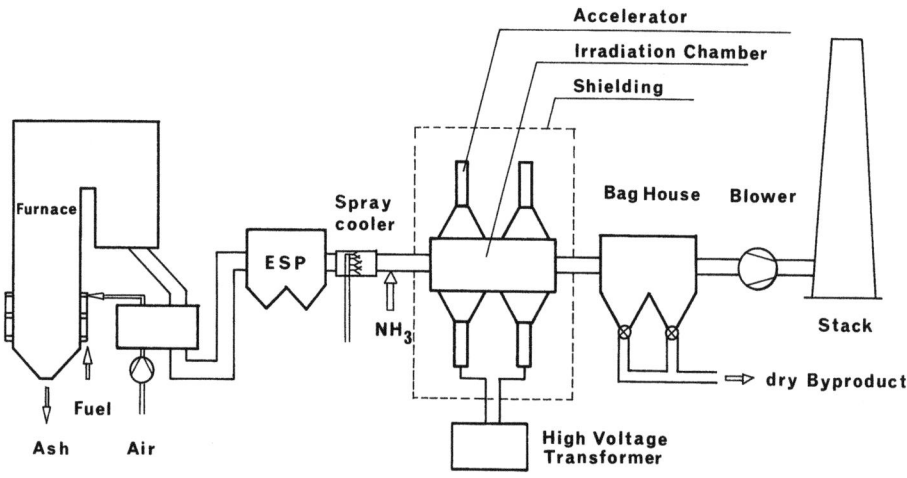

Fig. 1: Simplified Flow Chart of the EBDS Process for the Simultaneous Removal of SO_2 and NO_x from Flue Gas

In the first part of this paper an overview of the present knowledge regarding the chemistry and physics of the process will be given. Secondly the filtration of the product and the question of the product useability will be discussed. An estimation of the process costs concludes the paper.

2. Gas phase chemistry and removal efficiencies

2.1. PARAMETER DEPENDENCIES OF THE REMOVAL EFFICIENCIES

The main parameter determining the NO_x removal yield is the absorbed dose (fig. 2). Whereas the NO_x removal (%) increases with dose, the specific NO_x removals (ppm/kGy) decrease. Computer modeling studies suggest, that this is due to back reactions of products such as NO_2 and HNO_2 which readily decompose into NO, if their concentrations exceed certain limits (see also fig. 5) [10].

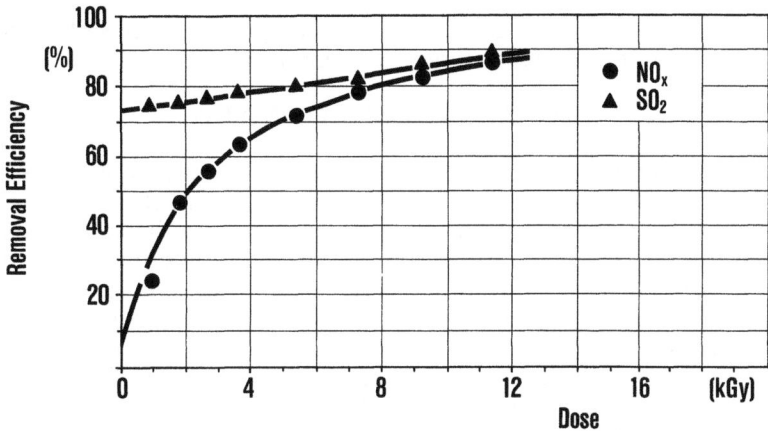

Fig. 2: NO$_x$ and SO$_2$ Removal Efficiencies as a Function of the Absorbed Radiation Dose (NH$_3$ Stoichiometry f = 0.98; rel. Humidity = 60 %; [NO$_x$]° = 205 ppmv, [SO$_2$]° = 350 ppmv).

Besides dose dependence, the NO$_x$ removals are influenced by the inlet concentrations of SO$_2$ and NO$_x$. Fig. 3 shows the increase of NOx removals as a function of the SO$_2$ inlet concentration. The reason for this experimental finding is, that the gas phase oxidation of SO$_2$ by OH radicals and oxygen produces O$_2$H radicals besides H$_2$SO$_4$. Subsequently, O$_2$H reacts with NO selectively and re-produces OH radicals. Therefore SO$_2$ can be regarded as some sort of catalyst for the NO$_x$ oxidation.

From the above data it is concluded that the EBDS process is especially suitable for power plants burning high sulfur coal and thus emitting high concentrations of SO$_2$. The enhancement of NO$_x$ removals by high SO$_2$ inlet concentrations also emphasizes the necessity to operate the EBDS process simultaneously.

The SO$_2$ removal efficiencies in the EBDS process are determined considerably by the parameters relative flue gas humidity and NH$_3$ stoichiometry (fig. 4). The NH$_3$ stoichiometry (f = [NH$_3$]°/ (2[SO$_2$]° + [NO$_x$]°)) is limited by the fact that certain emission levels of NH$_3$ have to be observed due to economical and environmental reasons. In West Germany an emission standard of 10 mg/m³ is presently under discussion. Therefore, the process is usually operated at NH$_3$-stoichiometries below one (e.g. 0.8 - 0.9). The parameter relative flue gas humidity determines the extent of the thermal reaction of SO$_2$ and NH$_3$, which occurs already without irradiation (see fig. 2). This reaction occurs mainly at the filter surface, but also in the ducts. The formation of deposits can be minimized, but not completely excluded, by maintaining the wall temperature of the ducts above 80 °C. The radiation induced oxidation of SO$_2$ is not a major pathway of the SO$_2$ removal in the

EBDS process, when compared to the thermal SO_2/NH_3 reaction. Experimental and theoretical data suggest that this pathway is responsible for approximately 10-30 % of the total observed SO_2 removal [4].

2.2. COMPUTER MODELLING OF THE GAS PHASE CHEMISTRY

Besides experimental studies computer models of the process chemistry are a very useful tool for a detailed investigation of the rather complicated reaction sequences which occur by the irradiation of flue gas with accelerated electrons. Kinetic models, which are based on the experimental data for atmospheric reactions have been developed by several authors [10-13].

The irradiation of flue gas by accelerated electrons causes excitation, ionization and fission of the gas molecules. Since the interaction of fast electrons with the gas molecules proceeds according to the mass fraction of each gas species, the direct interaction of electrons with SO_2 and NO_x is negligible. Energy is mainly transferred into N_2, O_2 and CO_2 molecules of the flue gas. The absorbed energy is given as a product of dose rate and residence time in the irradiation field.

According to the model calculations, the important gas phase reactions of the EBDS process are as follows:
- By radiolysis and charge transfer ions (N_2^+, O_2^+, H_2O^+) and excited molecules (N_2^*, O_2^*, H_2O^*, CO_2^*) are formed. These species react within microseconds with water vapour and oxygen and form radicals such as OH,

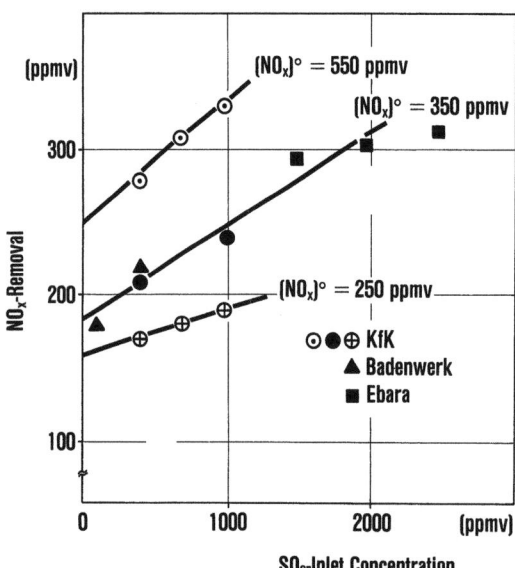

Fig. 3: NO_x-Removals (in ppmv) as a Function of SO_2 Inlet Concentration, with NO_x Inlet Concentration as Parameter; Dose = 15 kGy (KfK, Badenwerk) and 19 kGy (EBARA); NH_3 Stoichiometry = 0.9 - 1.1 (Data from KfK see Ref. [5], Badenwerk see Ref. [6], EBARA see Ref. [7].).

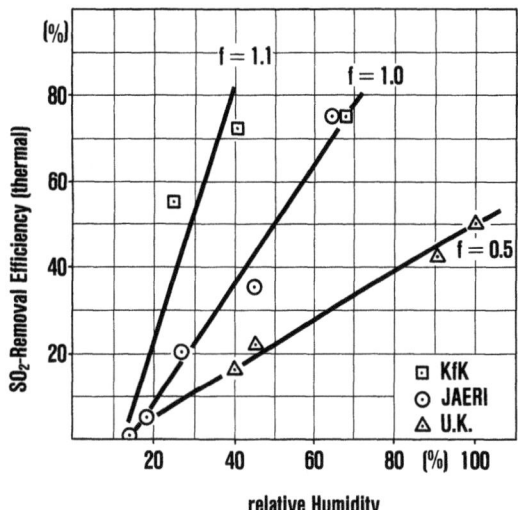

Fig. 4: Thermal SO_2 Removal Efficiencies as a Function of the Relative Humidity with NH_3 Stoichiometry as Parameter (Data from JAERI see Ref. [8], U.K. (University of Karlsruhe) see Ref. [9]).

O_2H, N, O, etc. The concentration of water vapour and oxygen is several magnitudes higher than the concentrations of NOx and SO_2. Therefore the ions react with water and oxygen only.
- From smog chamber studies and atmospheric measurements it is well known, that hydroxyl radicals (OH), oxygen atoms (O), hydroperoxy radicals (O_2H) and nitrogen atoms (N) are very short-lived species, which react fast with NO, NO_2 and SO_2. By these reactions OH, O and O_2H oxidize the NO molecules. The nitrogen atom transforms NO into molecular nitrogen. NO_2 is oxidized by OH and O_2H radicals, which leads to the formation of nitric acid (HNO_3). The oxygen atoms reduce NO_2 and form NO. This and other back reactions decrease the energy efficiency of the process [13]. SO_2 is exclusively oxidized by OH radicals. SO_3 which is formed in this reaction is hydrolized by water to sulfuric acid. If NH_3 is added to the flue gas before the irradiation, this gas reacts with OH radicals, forming NH_2 radicals. This radical may similarly as the N-atom reduce NO and NO_2. In these reactions molecular nitrogen and nitrous oxide are formed (see also section 2.4.1).
- After the formation of nitric acid (HNO_3) and sulfuric acid (H_2SO_4) these are neutralized by ammonia. The process of aerosol formation will be discussed in the following section.

2.3. AEROSOL FORMATION

The aerosol formation in the EBDS process is initiated by the nucleation of H_2SO_4, which has a very low vapour pressure under the usual operating conditions of the process. The aerosol is subsequently neutralized by NH_3 and undergoes changes in size distribution by coagulation and condensation. Computer modeling has been successfully applied to describe the formation of the H_2SO_4 aerosol [14]. Experimental studies have shown that under the usual

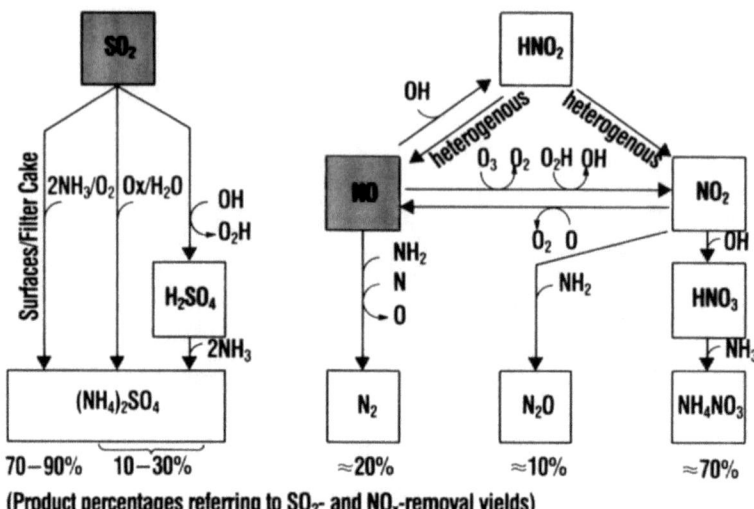

Fig. 5: Major Reaction Pathways of the EBDS Process (Data from KfK [4, 5] and JAERI [8, 18a, b].

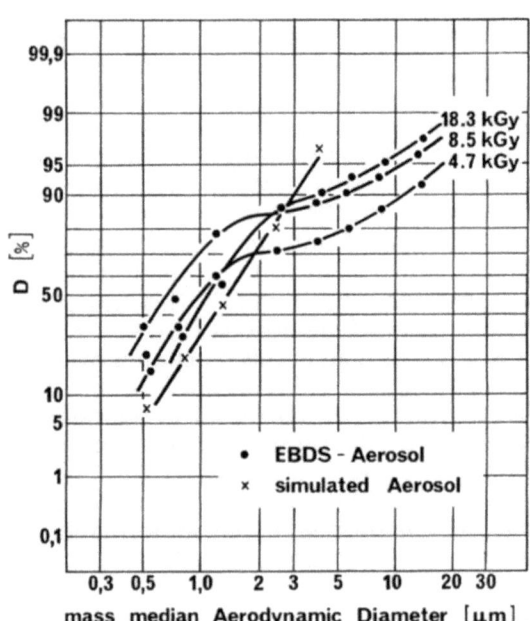

Fig. 6: Particle Size Distribution of the EBDS Aerosol [23] and Simulated Ammonium Sulfate Aerosol

operating conditions the EBDS aerosol has mass median diameters of less than 1 μm (fig. 6). The mass median diameters change with dose and flue gas humidity [15-16]. Besides purely physical aerosol processes, heterogeneous chemical reactions have been observed in the experiments. These reactions increase the amount of ammonium sulfate, as formed by gas phase reactions [4]. Computer modeling studies further suggest that heterogeneous decomposition reactions of HNO_2 play an important part in "back" reactions, which limit the energy efficiency of the EBDS process (see fig. 5) [10].

The aerosol mainly consists of ammonium sulfate with minor amounts of ammonium nitrate. The chemical composition of the aerosol depends on the removal yields of SO_2 and NO_x, dose and relative flue gas humidity [4, 17].

2.4. MATERIAL BALANCE

Further studies regarding the aerosol formation of the EBDS process showed, that the amount of solid products sampled from the irradiated flue gas was significantly lower than calculated from the SO_2 and NO_x removals. Therefore an experimental material balance for the N and S species was established.

2.4.1. Material balance of nitrogen species. As discussed above (see fig. 5) NO_x is transformed into ammonium nitrate and also molecular nitrogen and minor amounts of nitrous oxides. Whereas ammonium nitrate and N_2O are quite easily to be quantified, the source of molecular nitrogen is difficult to detect. Namba et al. [18a, b] performed experiments in a small flow system using ^{15}N labeled NO. Therefore it was possible to detect $^{29}N_2$ by mass spectroscopic analysis. Table 1 compiles the experimental and theoretical results regarding

Tab. 1: Experimental and Theoretical Balance of N-species

Species	Experiment 1[1]	Model	Experiment 2[2]	Model
a) in percent of removed NO_x				
NO_3^-[3]	94.8	67.1	52.2	36.4
N_2	18.5	38.8	n.d.	36.4
N_2O	17.3	19.4	23.1	27.3
Sum	131.6	119.7	75.3	100.0
b) in percent of inlet ^{15}NO				
^{15}N products	103.0	100.0	n.d.	100.0
^{14}N products	38.4	58.4	n.d.	n.d.

[1] Experimental conditions (taken from ref. [18]): SO_2 = 250 ppmv, ^{15}NO = 250 ppmv, NH_3 = 690 ppmv, O_2 = 18 Vol %, H_2O = 10 Vol %, Dose = 14 kGy
[2] Experimental Conditions (taken from ref. [19]): SO_2 = 888 ppmv, NO_x = 272 ppmv, NH_3 = 1190 ppmv, O_2 = 9 Vol %, H_2O = 15.2 Vol %, Dose = 12.5 kGy
[3] Nitrate determined from the product collected by electrostatic precipitator (ESP) [18] and by bag filter[19]; n.d. = not determined.

this experiment [19]. Two types of product species were detected and measured quantitatively, those containing the ^{15}N isotope, which originate from ^{15}NO, and those containing the ^{14}N isotope, which originate from input ammonia and molecular nitrogen. The material balance of the ^{15}N species is complete (103 %). The ^{14}N products amount to 38.4 % with respect to the inlet NO. The computer simulation of experiment 1 also predicts a significant amount of ^{14}N products (58.4 %), formed by the radiolysis of $^{28}N_2$ and $^{14}NH_3$.

The second experiment was performed at the AGATE pilot plant of KfK. In accordance with the prediction of the computer model the sum of the oxidized N-product (NO_3^- and N_2O) is less than 100 % (75.3 %). Nitrate was formed as an aerosol (29 % and by heterogeneous reactions in the filter cake of the bag filter (23.2 %). The latter fraction was determined from the difference between aerosol nitrate and the total amount of nitrate, as recovered from the bag filter. In contrary to the experimental and theoretical results in experiment 1, the sum of N-containing products in the model calculations does not exceed 100 %. The rather good agreement between the calculated and the theoretical results in the second experiment supports the conclusion, that the missing amount of NO (24.7 %) is due to the formation of molecular nitrogen by reducing reactions involving NO, NH_2 radicals and N-atoms [17].

2.4.2. *Material balance of Sulfur-Species.* Data from the AGATE pilot plant (experiment 2) show, that only 25.9 % of the sulfate mass was found in the aerosol. Significant SO_2 and NO_x removals were detected across the bagfilter. These account for 44.8 % of the SO_2 removal and for 23.1 % of the NO_x removals (see above). In addition, it was found that deposits were formed in the reaction chamber and the ducts, which were not quantified. By wet chemical and spectroscopic methods no other gaseous products were detected in the cleaned gas. Therefore the amount of sulfate in the deposits was estimated to be 29.3 % of the total amount of sulfate.

Tab. 2: Material Balance of Sulfate (in %)

Fraction	Experiment 1[1)]	Experiment 2[1)]
Aerosol	-	25.9
Bag Filter	-	44.8
ESP	47	-
Deposits	53	n.d.
Sum	100	70.7

[1)] Experimental conditions: see table 1

Data from experiment 1 show that 47 % of sulfate was collected in the electrostatic precipitator. The amount is somewhat lower comparing with the corresponding amount in the combined aerosol and filter fractions in experiment 2. This is due to the different irradiation conditions and the lower SO_2-concentration in the experiment. Half of the sulfate was detected in the irradiation vessel and in the pipes of the flow system as deposits, which were dissolved in water and analyzed. The sum of the sulfate fractions amounts to 100 %.

2.4.3. Aerosol Physics and Deposits. Size measurement of the EBDS-aerosol [15] and theoretical calculations [14] show that the mass median diameters are below 1 µm for most cases. For the AGATE pilot plant the extent of particle deposition has been calculated explicitly. The following deposition mechanisms were considered for turbulent gas flow: deposition by Brownian diffusion, by inertial effects and in bends [20]. According to this calculation, only the size fraction $d < 0.02$ µm is deposited at the duct walls by Brownian diffusion. Since less than 2 % of the aerosol mass are contained in this size fraction, the formation of deposits (see Tab. 2) is due to other (probably chemical) mechanisms. This is in accordance with the experimental result, that the chemical composition of the deposits does not correlate to the aerosol composition. The major component of the deposits is ammonium sulfate (99 weight-%) with very little ammonium nitrate (1 weight-%).

3. Filtration of the EBDS Aerosol

One of the major advantages of the EBDS process is, that it is a dry process. The aerosol should therefore be removed from the flue gas by dry filtration methods e.g. fabric filters or electrostatic precipitators (ESP). The mass concentrations of the aerosol in the irradiated flue gas are between 500 to 1500 mg/m^3. The mass median diameters of the rather hygroscopic aerosol are below 1 µm. In addition it had been found (see section 2.4.) that significant amounts of SO_2 and NH_3 are removed across the bag filter. The filtration of the aerosol is performed at high water contents and rather low flue gas temperatures. The combination of all these process features do not necessarily favour the dry filtration of the EBDS-aerosol.

Indeed, during the operation of larger EBDS pilot plants, difficulties were encountered regarding the operation of the fabric filters, because of a clogging of the filter bags. Frank et al. [21] therefore suggested to combine an electrostatic precipitator with a baghouse in series, in order to maintain stable operation conditions of the baghouse. Another approach to achieve improved performance of the baghouse is adding inert aerosol (e.g. fly ash, diatomaceous earth, lava dust) to the irradiated flue gas [22, 23]. By the third method the hygroscopic aerosol is diluted and the cleaning of the filterbag becomes more efficient. In the following the present state with respect to the filtration of the EBDS aerosol will be discussed in detail.

3.1. FILTRATION BY FABRIC FILTERS

Removal efficiencies: Submicron ammonium salt aerosols, released from production facilities, give rise to violet clouds. This is due to the light scattering properties of these aerosols. In order to prevent this, the maximum mass concentration of aerosol in cleaned offgas of an EBDS plant should be maintained below 10 mg/m^3. Table 3 lists the removal efficiencies for EBDS aerosol under a variety of experimental conditions such as filter media and face velocities.

The removal efficiencies were in the range between 99 up to 99.9 %

Tab. 3: Removal Efficiencies for the EBDS Aerosol

Run #	filter medium [1]	face velocity cm/sec	$m_{additiv}$ kg/h	c_m [2] mg/m³	c_∞ [2] mg/m³	η %
1	PTFEM	1.08	3.3	360 ± 48	1	99.8
2	PTFEM	1.08	2.1	554	2	99.7
3	PTFEM	1.45	3.0	799 ± 106	4	99.5
4	PTFEM	1.45	1.6	677 ± 3	8	98.7
5	PTFEM	1.45	0.9	1149	5	99.5
6	PTFEM	2.17	3.6	921 ± 53	2	99.8
7	PTFEM	2.18	2.4	723 ± 102	7	99.0
8	PTFEM	2.18	0.9	n.d.[3]	n.d.	n.d.
9	DT-NF	0.57	n.d.	1121 ± 77	< 1	<99.8
10	DT-NFA	1.08	0.9	517 ± 25	1	99.8
11	DT-NFPU	1.08	0.4	838 ± 90	1	99.9
12	DT-NF	1.08	0	897 ± 52	139	82-92

[1] PTFEM = PTFE-membrane on teflon fabric (PTFE/PTFE 280);
DT-NF = Dralon T needled felt (DT-DT 551);
DT-NFA = Dralon T needled felt with acid protection (DT-DT 551 + CS 17);
DT-NFPU = Dralon T needled felt with PU coating (GMD 4570-MF)
[2] mass concentration of EBDS-aerosol before (c_m) and after (c_∞) the bagfilter
[3] n.d. = not determined

regardless of the filter media used. Within experimental error no significant effect was found due to the face velocities nor the amount of added inert dust. From these results it can be concluded, that the removal of the aerosol occurs mainly on the filter cake, rather than on the fabric filter itself.

Cleaning efficiencies: The differential pressure of an aerosol filter may be expressed as the sum of the differential pressures of the cleaned filter and the filter cake. In the case of stable filter operation the differential pressure of the filter after cleaning (Δp_{min}) should stay constant. When the aerosol filter becomes clogged during the operation, the cleaning efficiency will decrease.

The experimental data show, that the cleaning efficiency in the filtration of the EBDS aerosol may be influenced by several parameters. As shown in fig. 7, the cleaning efficiency of the bag filter increases with an increasing amount of inert additives in the flue gas. Secondly the cleaning efficiency my also be improved by decreasing the face velocity. Thirdly the type of filter cleaning is important. Online cleaning (the filter is cleaned, while the gas is flowing) results in a rather low dedusting efficiency, after several hours of operation (see fig. 7). A much better behaviour was achieved, when the bag filter was cleaned offline (the filter was cleaned without gas flowing through it). By this method up to 50 hours of continuous operation were carried out without any

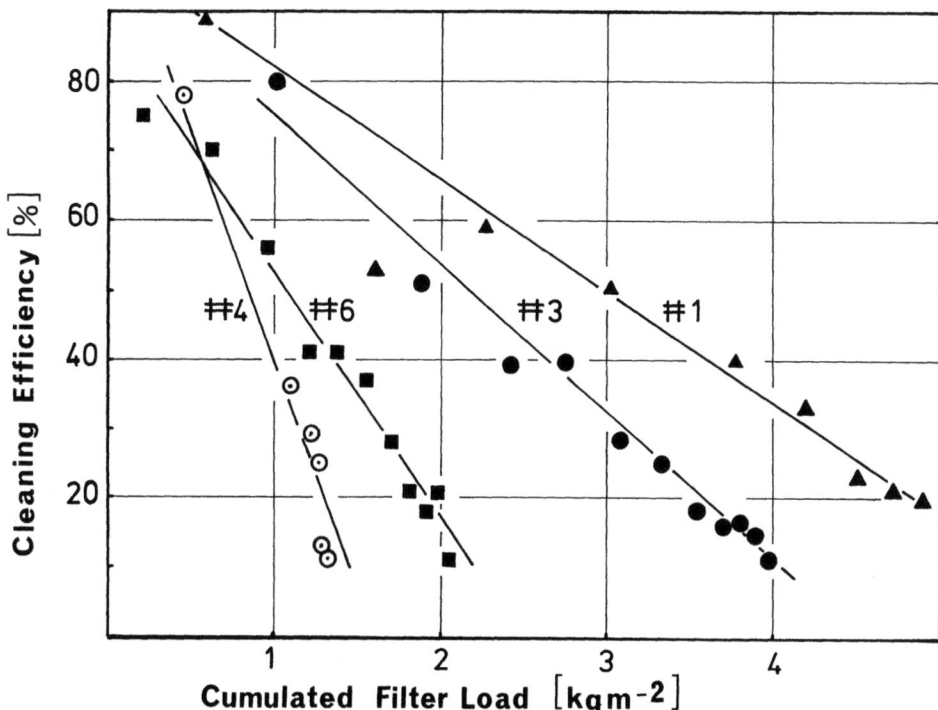

Fig. 7: Online-Cleaning: Pressure Differential of the Fabric Filter as a Function of the specific filter load (for exp. parameters see tab. 3).

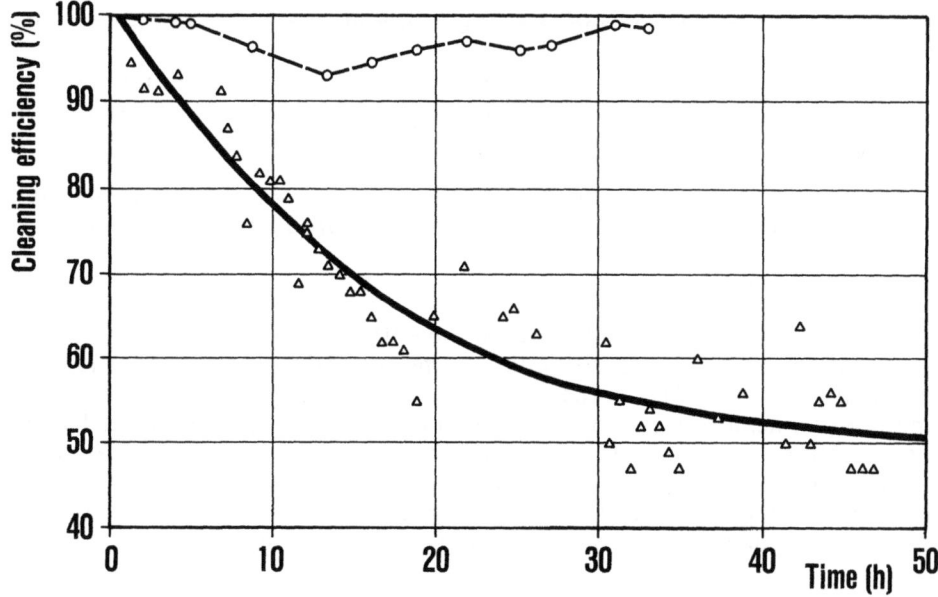

Fig. 8: Offline Cleaning (- - -) and Online Cleaning (——): Pressure Differential of the Fabric Filter as a Function of the Filtration Time

decrease in cleaning efficiency (see fig. 8). The cleaning efficiencies with offline cleaning were in the order of 95 %. Since in this method rather high amount of inert additives are required, thirdly online cleaning with recycling of the aerosol ladden additive was investigated. This method resulted in a stable cleaning efficiency also (see fig. 8). The cleaning efficiency in these experiments was lower than in the offline cleaning experiments. By this method the amount of inert additives was decreased substantially. The product obtained with online cleaning and recycling of the additive contained up to 25 % ammonium salts.

3.2. FILTRATION BY ELECTROSTATIC PRECIPITATORS (ESP)

Test runs for the filtration of the EBDS aerosol using ESP resulted in relatively high removal efficiencies (95 %). Other researchers [7] reported even higher removal efficiencies of 99 %. The product obtained from electrostatic precipitators does not contain additives. Therefore it can be used easily for the production of fertilizer.

Unlike as in the case of fabric filters, no additional reactions of SO_2 and ammonia take place in the ESP. Therefore the removal efficiencies for SO_2 are lower and the leakage of NH_3 in the process is higher. In order to achieve similar removal efficiencies for aerosols and gases, it is necessary to run ESP and fabric filter in series. The fabric filter is operated with rather high amounts of inert additives. By combining the products obtained from ESP and fabric filter rather high ammonium salt contents of about 90 % were obtained. These materials are suitable for the production of fertilizer.

3.3. OTHER FILTER TYPES

In addition to fabric filters and ESP other filters like gravel bed filters and wet ESP have been suggested.

Gravel bed filters have been shown to remove synthetic ammonium sulfate aerosol with high efficiencies. By sieving the aerosol ladden gravel after the filtration, high purity ammonium sulfate was obtained. The gravel can be recirculated into the filter [23].

The use of wet electrostatic precipitators produces solutions of ammonium salts. These solutions have to be dryed, in order to produce solid fertilizer.

3.4. CONCLUSION

Experiments regarding the filtration of the EBDS aerosols have shown that this part of the process requires a significant amount of development. The removal efficiencies of the EBDS aerosol are very high (99.9 %) for bag filters and methods for achieving a stable fabric filter operation have been found. Further optimization is still necessary in order to achieve higher concentrations of ammonium salts in the product. With regard to this aspect other filter types (ESP, wet ESP, gravel bed filter) are very promising. Nevertheless, these filter types have to be optimized also for this very special filtration task.

4. Product Useability

With respect to the long term application of the flue gas cleaning process, the question of product useability is of paramount importance. This is especially true for the EBDS process, because the product generated in this process is a very well soluble ammonium salt. This product can not be stored or disposed of in industrial dump sites. In the following, questions of market capacity for ammonium fertilizers and product characteristics will be discussed shortly.

4.1. MARKET FOR N-FERTILIZERS

The worldwide production for N-fertilizers was approximately 70 million tons in the year 1984/85. Within Europe about 5 - 6 million tons of ammonium sulfate-nitrate fertilizers are sold per year. The market share of this fertilizer type is mainly a question of the price. Due to the rapidly increasing desulfurization capacity for power plants it may be expected in midterm future, that sulfur deficiency of agricultural soils will become an important issue. Therefore an increasing use of sulfate containing fertilizers may be expected. A rough estimation shows that with a market share of 5 % of the total desulfurization capacity EBDS would supply approximately 5-10 % of the ammonium sulfate-nitrate sold per year in the FRG.

4.2. PRODUCT COMPOSITION

The composition of the EBDS product depends on several factors. Firstly it will be significantly influenced by the load conditions of the power plant and the removal efficiency of the ESP of the power plant, which precipitates fly ash. Secondly, the removal efficiencies and also the flue gas composition (coal composition) will also determine the composition of the product. Thirdly, the filtration method, which is used for the removal of the EBDS aerosol from the flue gas, is important.

With respect to the power plant, one has to consider, that the load conditions for coal fired plants are widely different in various countries. Whereas in the FRG coal fired power plants are mainly used to cover peak loads, coal fired plants in the United States will run in base load condition. Under peak load the flue gas composition is varying significantly, therefore the product obtained will also have a changeable composition. In order to obtain a product of sufficient purity (heavy metals, see below) is necessary, that the electrostatic precipitator, which removes the fly ash before the process, is in very good condition. With respect to the useability of the product it is important, not to contaminate the fertilizer with fly ash.

Depending on legislation and economics, the removal efficiencies for a EBDS plant will vary from country to country. The same is true for the coal composition. As discussed above, the EBDS process is especially suitable for high sulfur coal. Therefore it can be assumed, that the product obtained from this process will mainly consist of ammonium sulfate with smaller amounts of ammonium nitrate. If the coal has a high content of cloride, the EBDS product will also contain some ammonium-chloride.

In the case, that fabric filters are used for the removal of the EBDS aerosol, it is necessary to add an inert dust, in order to maintain a stable filter operation. To obtain a useful product, the amount of the additive should not exceed 10 -

20 %. The nature of the additive has to be chosen in such a way, that it will fit to the fertilizer production process. EBDS products, which are precipitated by ESP, usually contain no additive. They mainly consist of ammonium sulfate. This product is especially suitable for the production of fertilizers, since it contains high (above 20 %) amount of N.

One major concern with respect to the product useability has been the potential content of heavy metals in the EBDS product. Meanwhile numerous analyses have been performed with products from different EBDS pilot plants. All these analyses show, that the EBDS products does not contain significant amounts of poisonous heavy metals (see table 4).

Tab. 4: Typical Compositions of EBDS Products [1)]

Components	Sample 1 [2)] %	Sample 2 [3)] %	Sample 3 [4)] %
N	21.0	6.3	17.8
CaO	<0.05	3.6	6.3
MgO	<0.05	1.7	4.5
P_2O_5	n.d.	0.5	0.34
K_2O	n.d.	1.3	1.1
MgO	<0.05	1.7	4.5
Cl^-	n.d.	<0.1	0.79
SO_4^-	62 %	14.8	9.6
	ppm (w/w)	ppm (w/w)	ppm (w/w)
Cu	1.3	45	37
Zn	1.4	65	195
Pb	<1	4.8	3.2
Ca	<0.5	1.4	0.15
Cr	12	52	48
Ni	13	39	38
Hg	<0.01	<0.005	0.09
Fe	n.d.	4.1 %	3.6 %
Mn	n.d.	0.1	0.09

[1)] Samples 1 and 2 habe been prepared by irradiation of the gas from a heavy oil steam generator, sample 3 is from a coal fired power plant.
[2)] Sample 1 collected with ESP at the EBDS-Pilot Plant AGATE (KfK)
[3)] Sample 2 collected with Bag Filter at the EBDS Pilot Plant AGATE (KfK)
[4)] Sample 3 collected with Bag Filter at the EBDS Pilot Plant at RDK7 in Karlsruhe

Samples 2 and 3 contain inert additive.

Fertilizer tests were performed using gras and other crops (fig. 9). These experiments show that the EBDS product (sample 1 from tab. 4) is suitable for fertilization. No significant effects were detected, when the EBDS-product was compared to pure ammonium-sulfate fertilizer.

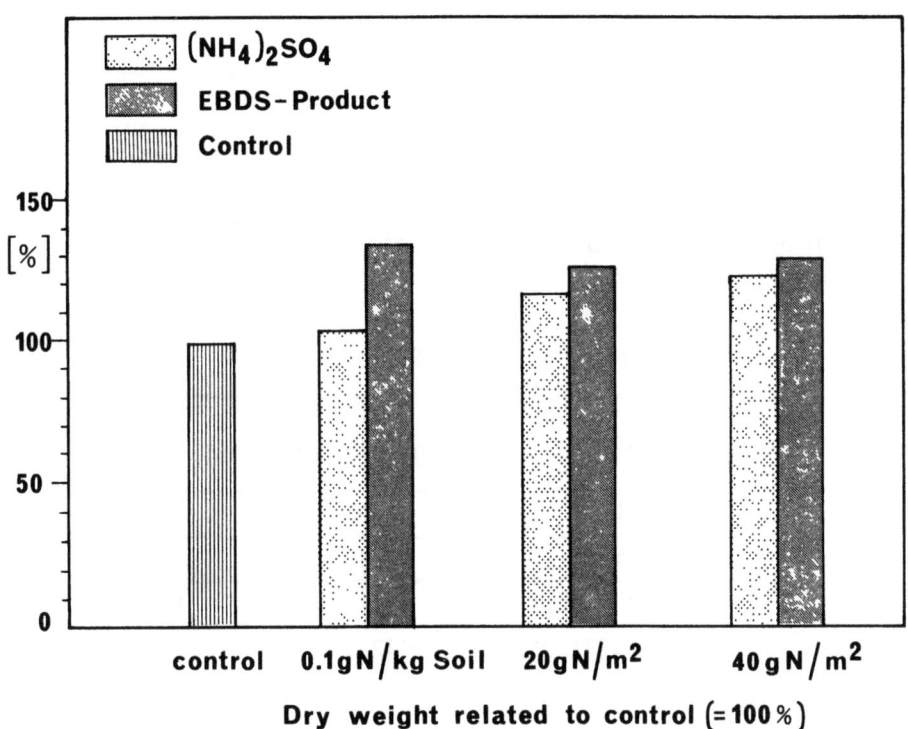

Fig. 9: Fertilizer tests with EBDS Product in Comparison with pure Ammonium Sulfate

4.3. CONCLUSION

The market for N-fertilizers in West Germany and also worldwide is large enough to accomodate a significant amount of the EBDS product. The product composition depends on several factors, and it can be expected, that it will be changeable. Therefore this product will certainly be only a raw material for the fertilizer production, rather than for immediate use. Nevertheless, the product is of sufficient purity in order to be used for agricultural purposes.

5. Economics of the EBDS Process / Estimation of Process Costs

Several simultaneous flue gas cleaning processes, among them the EBDS process, are now in the status of development or in the status of technical improvement. The decision of an utility for a certain process depends critically on the local situation. Especially for smaller and medium sized power plants criteria have to be considered like requirement for operating personnel, handling and storage of the reaction media, long term utilization or disposal of the reaction products, questions of waste waters, etc. Finally the investment and operating costs of the process under consideration must be comparable to other flue gas cleaning processes.

As pointed out above, the cost effectiveness of the EBDS process depends strongly on the inlet concentration of NO_x and SO_2. Thus the process seems to be most suitable for the simultaneous removal of NO_x concentrations below 400 ppmv and SO_2 concentrations above 1000 ppmv. With this flue gas composition, the necessary dose for obtaining removal efficiencies of 80 - 90 % for SO_2 and 70 - 80 % for NO_x is about 15 kGy. Assuming an accelerator efficiency of 75 %, this dose corresponds to 2.7 % of the electrical output of the power plant (see tab. 5).

The investment and operating costs of the EBDS process have been estimated by several researchers. The data from these sources are compared in table 5:

Tab. 5: Economics of the EBDS process

Source [1]	Investment ($/kW)	Operation (mill/kWh)
EBARA	209	15.9
DOE	334	17.6
KfK	290	12.3

1) EBARA [7], DOE [24], KfK [25]

According to these sources the investment costs range between 200 and 300 $/kW and operation costs are in the range of 15.9 - 12.3 mill./kWh. Considering the fact, that the EBDS process will be used especially for high sulfur applications, this cost range is reasonable, when compared to a ready developped technology such as the Wellmann-Lord-process. For this process estimates show investment costs between 350 - 360 $/kW and operating costs between 30 - 40 mills/kWh [26].

Concerning the investment and operating costs of the EBDS process, the accelerators have an exceptional position. Earlier studies on the industrial application on the EBDS process showed, that at the present price level may cause up to 50 % of the total investment. Therefore it is expected, that the EBDS process will be even more competitive in terms of cost if the accelerator prices will drop due to the mass production. Starting from the mid seventies, the specific accelerator price (DM/kW) has dropped almost by a factor of 2 - 3. This is due to the development of large industrial machines (500 - 600 kW).

In a recent estimation by Frank et al. [27] the specific investment cost of a EBDS plant was determined to be 200 $/kW. Using this value, we have calculated the operating costs for a 100 MW_{el}-EBDS plant. The data for this calculation are compared in table 6. Figures 10 and 11 give a graphic overview of these data.

Tab. 6: Estimated Operating Costs for an 100 MW$_{el}$- EBDS plant

	Unit	Amount	spec. Price	Unit	Sum	Unit
Operation cost						
Accelerator	kWh	2663	0.15	DM/kWh	400	DM/h
Blower	kWh	702	0.15	DM/kWh	105	DM/h
Other	kWh	1050	0.15	DM/kWh	158	DM/h
Water	m³/h	24.4	2.50	DM/m3	61	DM/h
Ammonia	kg/h	563.7	0.5	DM/kg	282	DM/h
Additive	kg/h	389	0.33	DM/kg	126	DM/h
Sum					1132	DM/h
	Load hours:	2000	4000	6000	8000	
Total cost	DM/h	1132	1132	1132	1132	
Maintenance (4 %)	DM/h	710	355	237	178	
Depreciation (7.5 %)	DM/h	1132	666	444	333	
Interest (8 %)	DM/h	1421	711	474	355	
Other (3 %)	DM/h	533	266	178	133	
Sum	DM/h	5128	3130	2465	2131	
Specific cost	Pfg/kWh	5.13	3.13	2.46	2.13	
(1 $ = DM 1.75)	mills/kWh	29.3	17.9	14.1	12.2	

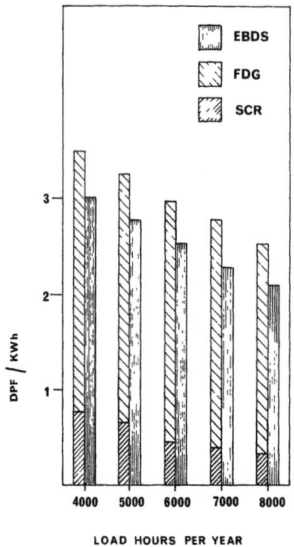

Fig. 10: Comparison of Cost for EBDS and for a SCR/wet FGD Combination as a Function of the Annual Load Hours

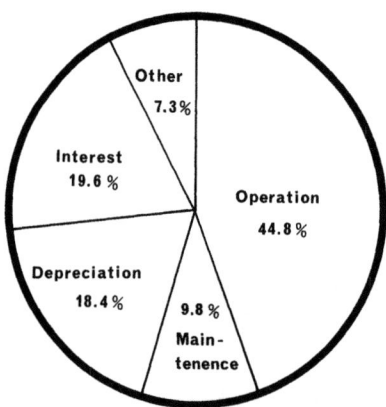

Fig. 11: Estimated Distribution of Operating Costs in a 100 MW$_{el}$-EBDS Plant (6000 Load Hours; see also Tab. 6)

6. Conclusion

Due to the interesting features of the EBDS process - both economically and technologically - a rapid international development is presently taking place. Pilot plants have been constructed and operated in the FRG, in Japan and in the U.S.A. Pilot units are also under construction in Poland and in China. For these countries the EBDS process is especially interesting, because they make use abundant of high sulfur coal. 100 MW demonstration units have been proposed in the U.S.A. and in Japan.

The up to date results show that the EBDS technology is safe and competitive. Further basic R and D work seems still necessary with respect to the thermal SO_2 reaction mechanisms and the collection of the submicron aerosol. Also a further evaluation of the usefulness of the fertilizer seems to be mandatory at this stage.

Acknowledgement

The author thanks Dr. H. Mätzing, Dr. H. Namba, Dr. S. Jordan (†) and Prof. Dr. W. Schikarski for stimulating discussions. The able technical assistance of Mrs. E. Schumacher and Mrs. B. Mathes is gratefully acknowledged. The fertilizer tests were carried out at KfK-HS/R by Dr. W. Schmidt.

References

1a. Jordan, S., Paur, H.-R., and Schikarski, W. (1988) Physik in unserer Zeit 19, 8-16

1b. Jordan, S. and Paur, H.-R. (1988) Beta-Gamma 1, 19-23

2. Jordan, S. (1988) Radiat. Phys. Chem. 31, 21-28

3. Tokunaga, O. and Suzuki, N. (1984) Radiat. Phys. Chem. 24, 145-165

4. Paur, H.-R. and Jordan, S. (1988) Radiat. Phys. Chem. 31, 9-13

5. Paur, H.-R. and Jordan, S. (1989) J. Aerosol Sci. 19, 1397-1400

6. Fuchs, P., Roth, B., Schwing, U., and Angele, H. (1988) Radiat. Phys. Chem. 31, 45-56

7. Frank, N. and Kawamura, K. (1988) Testing conducted on the EBARA E-Beam flue gas treatment system process demonstration unit at Indianapolis, Indiana; Final Report, DOE Contract # AE22-83PC60259

8. Machi, S., Namba, H. and Suzuki, N. (1987) IAEA-TECDOC-428, Electron Beam Processing of Combustion Flue Gases, IAEA, Vienna

9. Wittig, S., Spiegel, G., Platzer, K.-H. and Willibald, U. (1988) Kernforschungszentrum Karlsruhe, KfK-PEF 45

10. Mätzing, H. (1989) Kernforschungszentrum Karlsruhe, KfK 4494

11. Busi, F., D'Angelantonio, M., Mulazzani, Q.G., Raffaelli, V. and Tubertini, O. (1985), J. Radiat. Phys. Chem. 25, 47-55

12. Person, C.P., Ham, D.O. and Boni, A.A. (1985) Final Report for a Unified Projection of the Performance and Economics of Radiation-Initiated NO_x/SO_x Emission Control Technologies, Contract No. DE-AC 22-84 OC 70259, US Department of Energy, Pittsburgh Energy Technology Center, Pittsburgh, PA

13. Mätzing, H. (1989) Radiat. Phys. Chem. 33, 81-84

14. Mätzing, H., Paur, H.-R. and Bunz, H. (1989) J. Aerosol Sci. 19, 883-6, (1989)

15. Jordan, S., Paur, H.-R., Cherdron, W., and Lindner, W. (1986) J. Aerosol Sci. 17, 669-675

16. Paur, H.-R., Jordan, S., Baumann, W., Cherdron, W., Lindner, W., and Wiens, H. (1986) "Aerosols, Formation and Reactivity", Proceedings of the 2nd Int. Aerosol Conference, Berlin, 1024-1028, Pergamon, Oxford

17. Paur, H.-R. and Jordan, S. (1989) J. Aerosol Sci. 20, 7-12

18.a Namba, H., Aoki, Y., Tokunaga, O., Suzuki, R., and Aoki, S. (1988) Chem. Lett. pp. 1465-1468

18.b Namba, H., Tokunaga, O., Suzuki R., and Aoki, S. (1990) Material balance of nitrogen and sulfur components in simulated flue gas treated by an electron beam. Appl. Radiat. Isot., in press

19. Paur, H.-R., Namba, H., Tokunaga, O., and Mätzing, H. (1990) Proceedings of the Third International Aerosol Conference, Sept. 24-27, 1990 (Kyoto, Japan), in press

20. Fissan, H. and Schwientek, G. (1987) TSI Journal of Particle Instrumentation 2, 3-10

21. Frank, N., Hirano, S. and Kawamura, K. (1988) Radiat. Phys. Chem. 31, 21-28

22. Paur, H.-R., Jordan, S., and Baumann W. (1988), J. Aerosol Sci. 19, 1397-1400

23. Jordan, S., Baumann, W., Lindner, W., Paur, H.-R. (1989) PARTEC - 1. European Symposium Separation of Particles from Gases (Löffler F. ed.), Nürnberg, pp. 93-102

24. Tischer, R. (1989) Electron Beam Technology Pilot Scale Tests in U.S.A., International Workshop on Electron Beam Treatment of Combustion Flue Gases, Tokio, Japan, March 29-31

25. Schikarski, W., Jordan, S., and Körner, H (1988), VDI-Berichte Nr. 667, 85-102

26. Clean Coal Use Technologies, (1985) Vol. II, p. 103, DOE Washington, U.S.A.

27. Frank, N. (1990) Private Communication

PRIMARY MEASURES FOR NO_x REDUCTION

H.G. BOS
Stork Boilers
P.O. Box 20
7550 CB Hengelo
The Netherlands

1 Introduction

There are numerous ways to reduce the NO_x emission of industrial and utility boilers. We can distinguish between so-called primary and secondary methods.

With primary methods we mean those methods which affect the combustion process. That is, by means of primary methods the combustion process is modified in such a way that less NO_x will exit the furnace. This is achieved without addition of any chemicals other than fuel, air and flue gas.

Secondary methods are those methods which clean the flue gas. That is NO_x will be removed from the flue gas without affecting the combustion process. Most often this is done by adding chemicals, other than fuel, air or flue gas, to the furnace or after completion of the combustion process to reduce NO to N_2, with or without using a catalist.

In this paper we will concentrate on primary methods. Chapter 2 gives an explanation of the chemistry of NO_x formation and reduction. In the next chapters the different primary technics of NO_x reduction will be described.

2 NO$_x$ chemistry

2.1 GENERAL

The expression NO$_x$ can be used to indicate a set of nitric oxides. Most important in the combustion process are NO and NO$_2$. Recently there has been a growing interest in and knowledge of N$_2$O. Still in the combination "NO$_x$" and "combustion" the meaning of NO$_x$ is confined to NO and NO$_2$. NO and NO$_2$ are important in respect of acid rain. N$_2$O influences the greenhouse effect.

For a clear understanding of the different methods by which we try to reduce the NO$_x$ emission it is necessary to be familiar with the basic principles of NO$_x$ formation and destruction.

Basically we distinguish between three ways in which NO$_x$ is formed.

a. thermal NO$_x$ formed by a direct reaction of gaseous N$_2$ and O$_2$

b. fuel NO$_x$ formed from chemically bound N in the fuel

c. prompt NO$_x$ formed from gaseous N$_2$ via intermediate products.

The oxidation of nitrogen proceeds through NO to NO$_2$. There is an equilibrium between the concentration of NO and NO$_2$. At ambient temperatures all NO will convert to NO$_2$. At high temperatures above 1000C the equilibrium is on the NO side. This means that at the high temperatures in the combustion process mainly NO is formed. When after completion of the combustion the flue gases are cooled some of the NO will oxidize to NO$_2$. This conversion proceeds slowly at the lower temperatures in the convection part of the boiler. Therefore most of the NO$_x$ leaving the stack is still NO. On the long term all NO in the atmosphere converts to NO$_2$. The ratio of NO and NO$_2$ in the flue gases leaving the stack depends on the fuel, the combustion process, the boiler design, etc. For example coal fired boilers and also gas turbines have a relatively high concentration of NO$_2$ in their flue gases.

2.2 THERMAL NO$_x$

Thermal NO is formed from gaseous N$_2$ and O$_2$ directly. That is without more or less complex intermediate compounds.

Reactions involved are:

$$O + N_2 \rightarrow NO + N$$
$$N + O_2 \rightarrow NO + O$$
$$N + OH \rightarrow NO + H$$

Which one of these reactions is most important depends on the relative concentrations of the compounds involved. E.g. the OH radical concentration is relatively high in fuel rich flames. The most important parameters in this reaction mechanism are the temperature and the residence time in the high temperature region. Therefore the name "thermal NO". The reaction rate is about exponential dependent on the temperature. When there are two flames with the same mean temperature the flame with the most flat temperature profile will produce the lowest NO_x level.

Below a certain minimum temperature there is hardly any NO_x formation. In the literature values of this minimum temperature can be found ranging from 850C to 1300C. Our own experience is that at a temperature of 1000C the thermal NO production still is minimal.

2.3 FUEL NO

The nitrogen of fuel NO originates from chemically bound N in the fuel. The transformation from fuel N to NO passes through a number of intermediate products like CHN, CN, NC, NCO, etc.

There is not yet a complete description of this process. However, the basic principles and the main factors of influence are well known. The process starts with pyrolysis of the fuel. In this step most of the N is converted to HCN. This compound can transform to NO, N_2 and/or NH_3. Several possibilities are shown in figure 1.

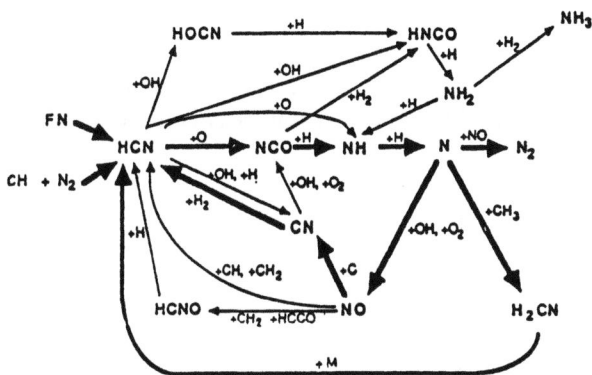

Figure 1: NO formation reaction mechanisms

In general the next statements can be done about fuel NO formation:
- The amount of NO produced depends on the N content of the fuel. However, this relationship is not linear. The fraction of N which converts to NO decreases with an increasing concentration of N.
- The conversion of N to NO is to a high degree independent of the temperature.
- The O_2 concentration is an important parameter.
- With coal we can distinguish between volatile matter and char combustion. The conversion of N in volatile to NO proceeds with a greater efficiency than from N in char to NO.

2.4 PROMPT NO

Prompt NO is formed out of gaseous N_2. The first step in this reaction mechanism is a reaction between N_2 and radicals like CH_2 and CH:

$CH_2 + N_2 \rightarrow HCN + NH$
$CH + N_2 \rightarrow HCN + N$

The reaction mechanisms involved can be studied by analysing flames of fuels without chemically bound N., i.e. gas flames. In oil and coal flames it is not well possible to distinguish between fuel and prompt NO formation.

Factors of influence in the prompt NO formation are:
- Concentration of O atoms (radicals)
- Excess air. I.e. in fuel rich flames the production of prompt NO is strongly promoted.
- The temperature is of minor influence.

2.5 N_2O

As mentioned before there is a growing interest in the N_2O emission. It is found that in certain situations the conditions that prevent NO formation or stimulate NO reduction also stimulate N_2O production.

Especially at relatively low temperatures the formation of N_2O is preferable to that of NO. Also reducing reagentia like NH_3 and CN_2H_2 can be converted to N_2O at relatively low temperatures.

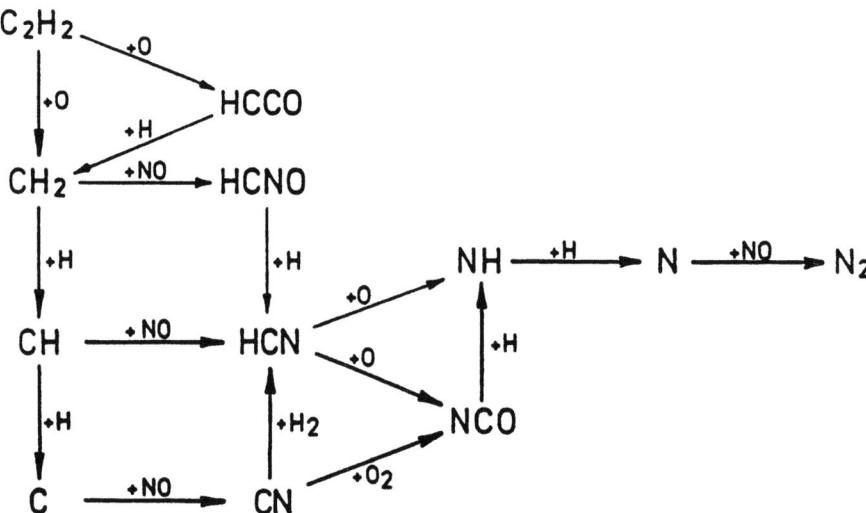

Figure 2: Some NO reduction reaction mechanisms

2.6 NO REDUCTION

From the secondary NO_x methods it is well known that NO can be reduced to N_2 by addition of components like NH_3 and CH_2N_2. The reduction proceeds through intermediates like HCN, CN and NH. These intermediate products are also available during the combustion process. This means that it should be possible to use fuel as a reducing agent.

From further analysis of this process it is clear now that using fuel as a reducing agent besides components like NH and CN radicals also CH_i radicals are very important in the NO reduction process. Some of the possible reaction mechanisms are shown in figure 2. The challenge nowadays is to modify the combustion process in such a way that the NO reduction is preferred to NO formation.

3 Primary methods of NO_x emission reduction

3.1 GENERAL

The primary methods are divided into two groups:
- modifications of burners
- modifications of the combustion system

The several low NO_x techniques concerning modifications of the combustion process will be explained using circular flow burners and the opposed firing system. The principles, however, also apply to different burners and combustion systems.

When reduction percentages are mentioned, the base situation is: single register burners and single stage combustion system, unless specified otherwise.

3.2 MODIFICATIONS OF BURNERS

The statements in this chapter basically apply to wall, roof or bottom fired furnaces. In these cases a burner can be treated individually in the first place, although there is an interaction between the flames. (see ch. 3.3).

With tangential firing this is just opposite: in the first place this combustion system shall be considered as one, furnace filling combustion process. It is, however, possible to distinguish individual burners and flames. The tangential combustion system has been more NO_x friendly than the other system for a long period of time. Nowadays the newest type of circular flow burners can reach lower NO_x emissions than the tangential firing system, at least for coal firing.

There are several types of low NO_x burners. These can be divided into two groups, called first and second generation. They differ basically in their design philosophy.

3.2.1 *First generation burners*. The main aim in the development of these burners was to suppress the NO_x formation. This is mainly achieved by delaying the combustion and splitting up the flame into a fuel rich and a fuel lean zone.

To reduce the formation of thermal NO_x it is necessary to reduce the flame temperature. To reduce the formation of fuel NO_x it is necessary to burn the fuel as much as is possible under fuel rich conditions.

The highest flame temperatures are reached with the combustion of a well mixed, (nearly) stoichiometric fuel-air mixture. Addition of more fuel or more air will cause a reduction in the combustion temperature. The addition of excess air is most effective for this temperature reduction. On the other hand, the addition of excess fuel will reduce the relative availability of oxygen and therefor also reduce the fuel NO formation.

The concept of the first generation low NO_x burners aims at the creation of fuel rich and

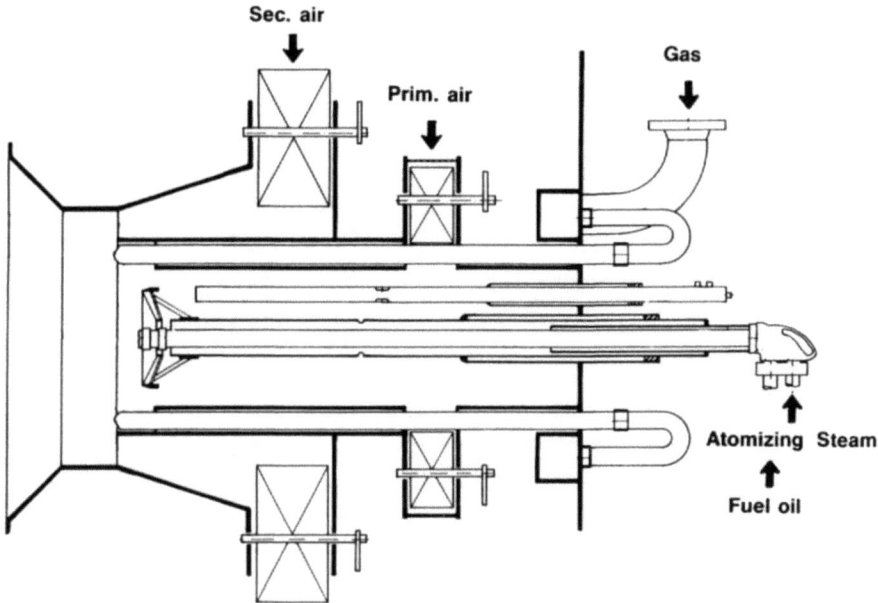

Figure 3: First generation Double Register Burner for gas and oil

fuel lean zones in the flame. This can be done either by control of the air fow pattern or by influencing the fuel injection or a combination of both

In the double register burner shown in figure 3 the air is divided into two layers, called primary air and secondary air. The fuel injection is designed in such a way that in the centre of the flame fuel rich conditions exist and in the outer layer fuel lean conditions. The completion of the mixing of fuel and air is therefor delayed. Part of the combustion proceeds under fuel rich conditions and the peak temperatures in the flame are reduced.

With burners based on the above-mentioned principles NO_x reduction levels up to 40% can be obtained.

Flue gas recirculation in fact belongs to the "modifications of the combustion process". (ch. 3.3.2). But it has also an impact on the burner design. There are two methods to add flue gas to a burner:
- mixed with the combustion air
- as a separate flow

With the first method the main effect is a reduction of the flame temperature. The second method is shown in figure 4. The flue gas is injected between the primary inner air and the secondary outer air, causing a better separation of the fuel rich and fuel lean zone.

In general low NO_x burners apply lower air velocities. Thus these burners require larger burner throats than conventional burners of the same heat imput. The slower combustion rate furthermore causes an increase in flame length.

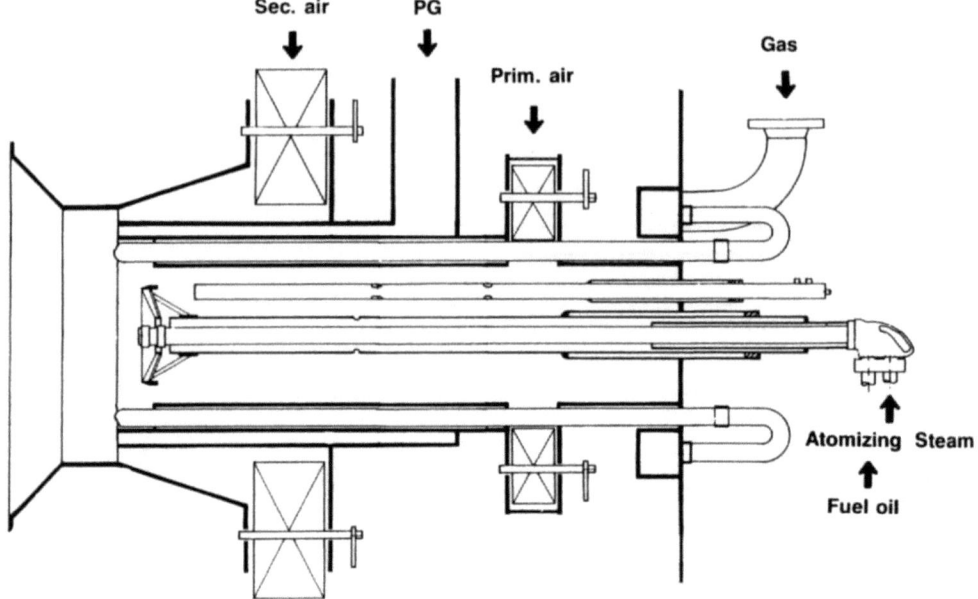

Figure 4: First generation Double Register low NO_x burner for gas and oil. With separate flue gas recirculation (PG)

3.2.2 Second generation burners. The main aim in the development of this burner type is to use NO_x reduction reactions in the flame. The first burner that uses this principle is the HTNR burner of Babcock-Hitachi (Japan). This is a coal fired burner. The combustion of this burner is illustrated in figure 5 and figure 6. The construction of this burner is shown in figure 5, the different zones in the flame in figure 6. In zone A, close to the burner are

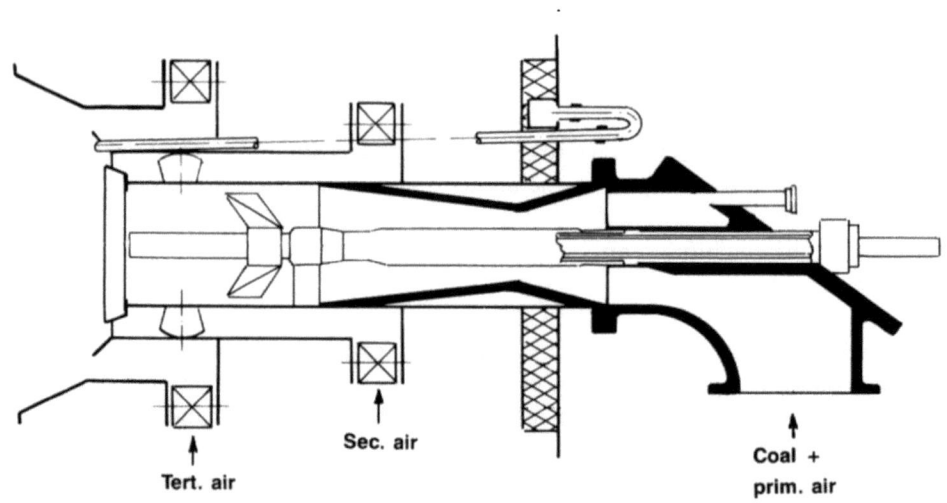

Figure 5: Second generation low NO_x coal and natural gas fired burner

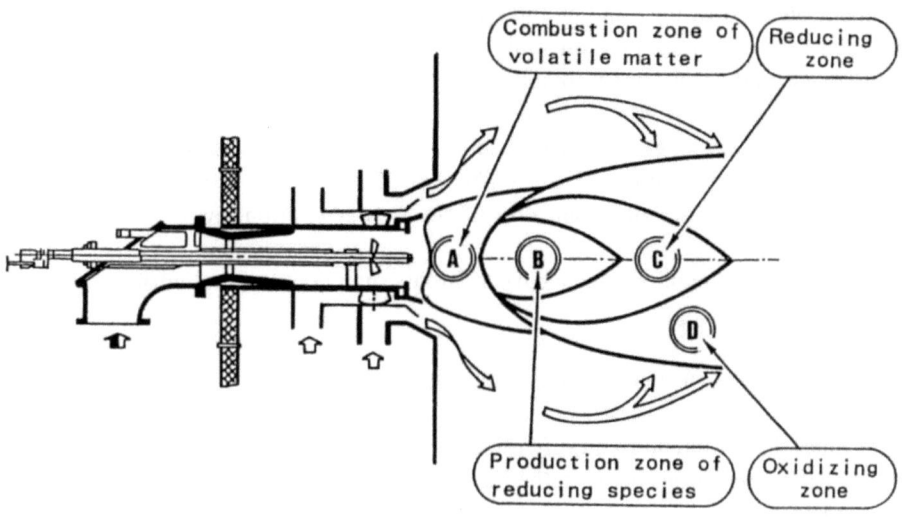

Figure 6: Coal fired second generation low NO_x burner. Different zones in the flame

very high temperatures to promote a fast devotalisation and initial combustion. In zone B, with a low O_2 concentration, reducing species are produced. The NO production in zone A is relatively high, but in zone C most of this NO is reduced to N_2.

After mixing with the outermost air layer (tertiary air) the combustion - of mainly the char - will be completed at relatively low temperatures.
The concentration of NO and some other components along the axis of the flame are shown in figure 7.

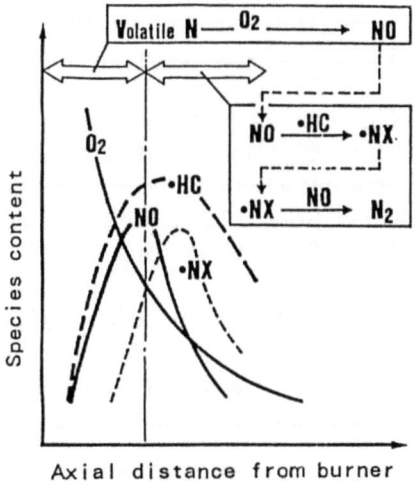

Figure 7: Concentration of different components along the axis of a second generation coal flame

Experiments on a test furnace and on full scale have shown that these burners also for natural gas reach very low NO_x emission values.

For coal firing and single stage combustion this burner can achieve a NO_x emission of about 600-700 mg/m_o^3 in existing furnaces, with the UBC <5%. In new boilers and in combination with two stage combustion (ch. 3.3.4) the NO_x emission level will be lower than 300 mg/m_o^3, again with the UBC <5%. For gas firing it is expected that this burner can achieve NO_x emissions of about 50% of the first generation low NO_x burner.

This burner type also requires larger burner throats than conventional burners, but there is no increase in flame length.

3.3 MODIFICATIONS OF THE COMBUSTION PROCESS

These technics do not intend to modify the combustion process of an individual burner, but to influence the interactions between the flames. In the next chapters the following methods will be explained:
- low burner zone heat release
- flue gas recirculation (F.G.R).
- off-stoichiometric combustion (O.S).
- two-stage combustion (T.S.C).
- in-furnace NO_x reduction (I.F.N.R).

3.3.1 *Low burner zone heat release.* By using a larger furnace and enlarging the distance between the burners, there is an increase in heat absorption to the furnace walls close to the burners. This causes a decrease in flame temperature and thus reduces the thermal NO_x formation.

3.3.2 *Flue gas recirculation.* This method is already mentioned in chapter 3.2.1. Flue gas can be added to the furnace in three different ways. The oldest method is through separate openings in the furnace. (e.g. in the furnace bottom). This method of flue gas recirculation is basically used for reheat steam temperature control, but also influences the NO_x production. The NO_x reduction effect is caused by decreasing the furnace temperature.

To achieve a large effect on NO_x formation peak temperatures should be lowered. The flue gas, therefore, should be available at the places with the highest temperatures, i.e. in the centre of the flame. This is only possible when the flue gases are mixed with the combustion air or added directly to the burner. (see ch. 3.2.1). Which one of these two methods is the most effective depends on the specific burner design.

The maximum amount of the flue gas which can be added to a burner depends on the flame stability. This is also influenced by the combustion air temperature. When the temperature of the air-flue gas mixture is about 300C the minimum O_2 level necessary for a stable combustion is about 16%. This corresponds with about 30% of flue gas recirculation.

Flue gas recirculation mainly suppresses the thermal NO_x formation. Therefore, this method is very useful for gas firing. The effect with oil firing is less. For coal firing this method is of no use. For gas firing a reduction of 60% or more can be achieved.

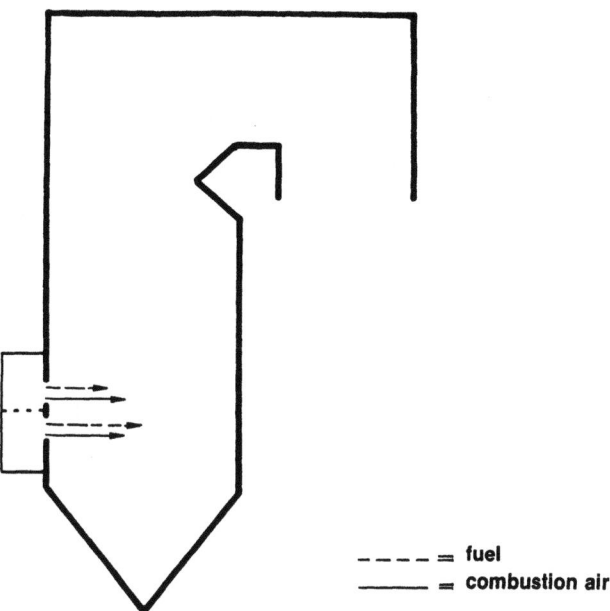

Figure 8: Off stoichiometric combustion system

3.3.3 *Off-stoichiometric combustion.* The off-stoichiometric combustion method can be applied only when the burner arrangement meets some constraints. This method is illustrated in figure 8 for a front fired furnace. There should at least be 2 levels of burners. The air and fuel are not equally distributed among the burner levels. The upper burners will get more air and/or less fuel than the lower burners. This causes a shift in the burner stoichiometry. The lower burners operate at fuel rich conditions ($\lambda \leq 1$), the upper burners at fuel lean conditions.

With this system the upper burners more or less operate like after-air ports (see ch. 3.3.4). So the burners should be arranged in such a way that a good mixing of the flames is achieved to get a complete combustion without substantial increase in excess air.

This method is mainly applied to existing installations. It is a relatively cheap and easy method to obtain a limited NO_x reduction of up to 25% for gas and oil firing. For coal firing it is less suitable. An increase in UBC can be expected.

3.3.4 *Two-stage combustion.* With this method all burners operate at fuel rich conditions. In principle all the burners maintain the same air to fuel ratio. The air necessary for a completion of the combustion is added to the furnace through after-air ports. (see figure 9). The fuel rich conditions at the burners cause lower flame temperatures and also a lower O_2 concentration. So the formation of thermal NO_x as well as of fuel NO_x is suppressed. The completion of the combustion, downstream of the after-air ports is at relatively low temperatures. In this phase of the combustion process fuel NO_x is of minor importance and thermal NO_x formation is effectively suppressed by the lower temperatures.

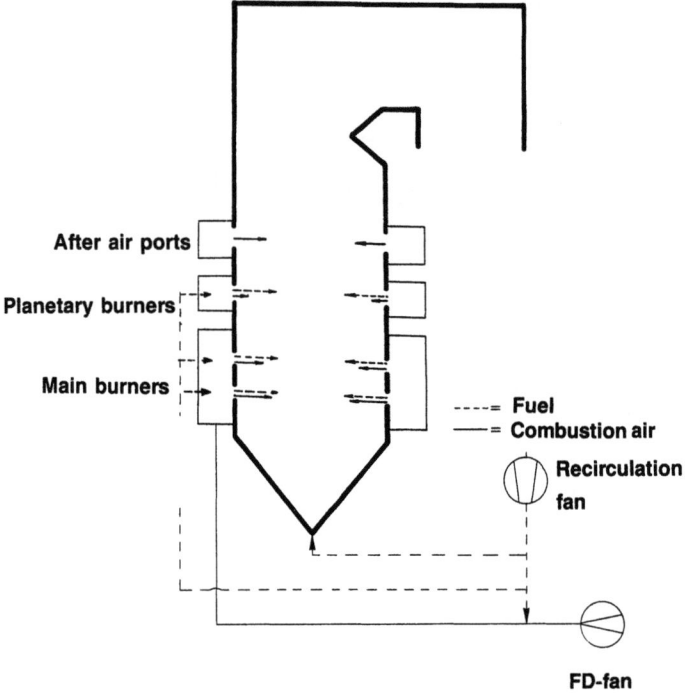

Figure 9: Two stage combustion system

To achieve a complete combustion at a normal excess air ratio very much attention should be paid to the mixing of the after-air with the combustion gases.

The effect of two-stage combustion on the NO_x emission depends on the air-fuel ratio at the burners. The lower this ratio, the lower the NO_x emission. However, at a certain limit there is a kind of saturation. The optimum value of λ for gas and oil firing is between 0.7 an 0.8. (see figure 10).

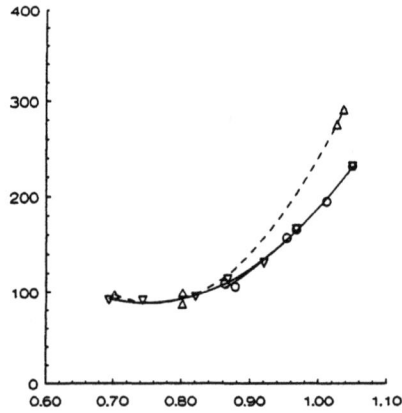

Figure 10: NO_x emission with two stage combustion as a function of the burner stoichiometry

For coal firing the situation is a bit different. The air to fuel ratio here is not limited by the effect on NO_x, but by the behaviour of the combustion. Combustion in a reducing atmosphere changes the slagging properties of coal. The ash melting temperature will shift to a lower temperature. The combination of reducing atmosphere, low ash melting point and sulphur may cause many troubles with furnace slagging and pipe corrosion. For this reason at coal firing the air to fuel ratio at the burners will be between 0.9 and 1.0.

Two-stage combustion is very effective in NO_x emmission reduction for gas, oil and coal firing. With gas and oil a reduction of 50% up to 60% can be achieved. Due to the higher stoichiometric ratio at the burners with coal firing, the effect with coal firing is limited to 30% - 40%.

3.3.5 *In-furnace NO_x reduction.* As explained in chapter 2.6 it is very well possible to convert NO to N_2 when the necessary conditions are met. This means: enough, but not too many reducing species (like CN, NH etc.), low O_2 level, sufficient high temperature and sufficient long residence time under those conditions.

The IFNR system described here was developed by Babcock-Hitachi. Stork Boilers applied this system in a demonstration project in the Flevo Power Station in the Netherlands.

With the IFNR process the furnace is divided into three zones (see figure 11):
1. The main combustion zone
2. The reduction zone
3. The combustion completion zone

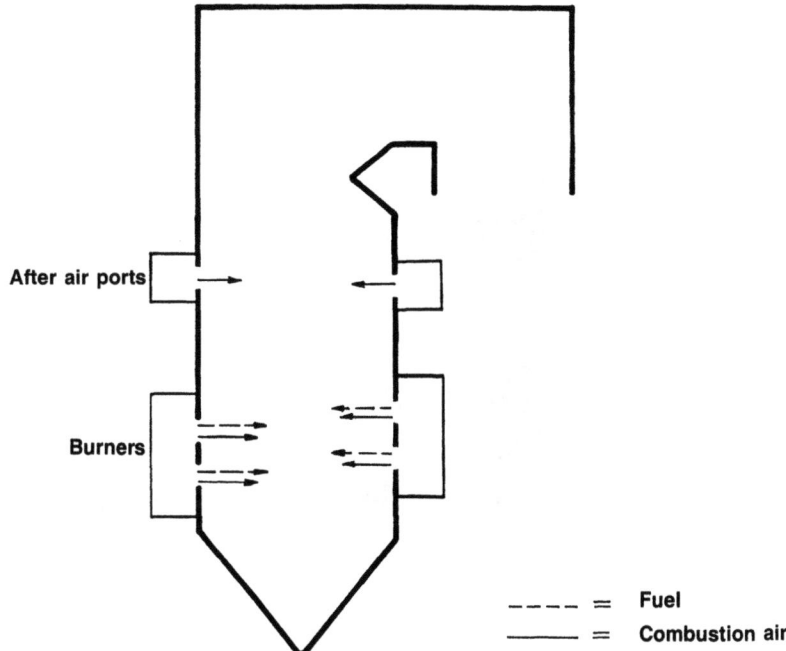

Figure 11: In-furnace NO_x-Reduction system for natural gas and oil firing

In the main combustion zone about 65 to 75% of the fuel is burned at fuel rich conditions. Also flue gas recirculation is applied to the burners to achieve the lowest possible NO_x concentration in this zone. In the second zone the rest of the fuel is added through planetary burners. These burners operate at a stoichiometric ratio of about 0.5. If the burners and the furnace are well designed a NO_x reduction in this zone of about 80% will be obtained. In the third zone the remaining air is injected through after-air ports. Some NO_x is formed in this region, due to the relatively high amount of N containing compounds leaving the reduction zone. These compounds are partially converted to NO. If the process is not an optimum design, this effect may be larger than the reduction in the second zone. The overall effect of NO_x reduction in the second and third zone is about 50%. (see figure 12).

Because of the necessary residence time in all three zones this combustion process needs a relatively large furnace. The system can be applied to gas and oil firing. For coal firing this system is also working. However, to achieve substantial reduction in the reduction zone very high volatile coals are needed. As an alternative oil or gas may be used as the reducing fuel. There also may arise problems with slagging due to the reducing atmosphere in the third zone.

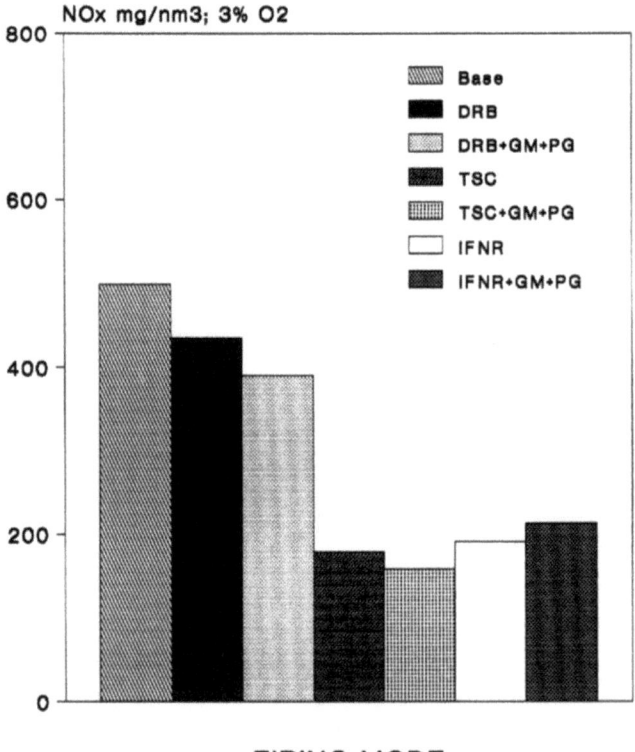

Figure 12: Field results with In-furnace NO_x-Reduction system gas firing

With the In-Furnace NO_x Reduction system, including flue gas recirculation, at the Flevo demonstration project an emission level of 44 mg/m$_o^3$ for gas firing was achieved. This is a very low value for a retrofitted boiler and a limited amount of flue gas recirculation. It is possible to achieve emissions lower than 20 mg/m$_o^3$ at steady state conditions with gas firing.

3.4 COMBINATION OF METHODS

All technics mentioned can be used in combination with each other. For a number of combinations the effects are cummulative. E.g. a certain amount of flue gas recirculation gives a reduction of 40% in combination with conventional burners and single-stage combustion. In combination with low NO_x burners or with two-stage combustion the reduction effect of the same amount of flue gas recirculation also will be approximately 40%. This rule, however, only can be applied to give a rough indication.

Which method or combination of methods should be used depends on the local conditions, e.g. existing or new boiler, environmental rules, fuel(s) to be fired etc. Generally speaking, the first method to be considered should be low NO_x burners. For the second method there is a choice between two-stage combustion and flue gas recirculation. For the combustion of N containing fuels (heavy oil, coal) two-stage combustion is most effective. For natural gas firing flue gas recirculation is also worthwhile. With the combination of low NO_x burners, two-stage combustion and flue gas recirculation for gas firing very low NO_x emission levels can be achieved.

Other combinations are of minor importance, keeping in mind that in-furnace NO_x reduction for gas and oil is always in combination with flue gas recirculation.

Off-stoichiometric combustion is most often used as a reduction method in those cases where the allowed emission levels are just not met.

For new installations also the furnace dimensions should be designed in accordance with the desired NO_x level.

State of the art at this moment is that with a combination of reduction methods the next figures can be obtained:

natural gas	< 50 mg/m$_o^3$
heavy oil	<150 mg/m$_o^3$
bituminous/subbituminous coal	<300 mg/m$_o^3$

3.5 SIDE EFFECTS

All NO_x emission reduction methods more or less influence the boiler operation. A careful consideration of these aspects should be performed before a decision is made in respect of the choice of the emission reduction method.

Some of the most important aspects are listed below:
- Low NO_x burners. These require larger burner throats than conventional burners. The first generation burners also have larger flames.
- Low burner zone heat release. This gives a larger furnace. There is a restriction on the furnace gas exit temperature, which should not be too low, especially with coal firing. For the second generation of low NO_x burners too low a furnace temperature may have a negative influence on the performance in respect of the NO_x reduction. Reduction of NO to N_2 requires a certain minimum temperature in the flame.
- Flue gas recirculation. By using this method the heat transfer to water and steam is changed considerably. Radiation will be less and the convective heat exchange will increase. For a retrofit this means that the convective heat section of the boiler should be modified. Because of the larger gas flow through the boiler there is an increase in fan power for the combustion air fan and/or the induced draught fan. There is also an extra power consumption because of the installation of a flue gas recirculation fan. New installations can be designed for this large amount of recirculation gas with limited additional costs.
- Two stage combustion. The mixing of the after air with the combustion gases is of main importance. If the system is well designed there is no influence on the boiler operation nor the boiler efficiency. In case of insufficient mixing and/or insufficient residence time between the after air ports and the furnace exit the excessair required to achieve a complete combustion will increase significantly. This extra air should be fed to the after airports and therefor hardly has any influence on the NO_x emission. The main effect of this increase in excess air is a decrease in boiler efficiency.

4   Further developments

There is a competition between primary and secondary methods. For new installations primary methods are more cost effective (see figure 13). For retrofits this depends on a number of local circumstances. However, here also primary methods in most cases will be more cost effective. Sometimes it will not be possible to achieve the required emission levels by primary methods only.

The further development in primary methods aims at low cost NO_x emission reduction.

For gas firing very low figures, <20 mg/m_o^3, can be achieved by combined combustion modifications. The development aims at lower NO_x emissions by burner modifications only. This will save on flue gas recirculation and thus on investment and operating costs (fan power).

For oil and coal firing lower emission values can be obtained. Here also the main investment is directed to the development of new burner technology. By improving the second generation low NO_x burner a further NO_x emission reduction of about 35% seems possible without any increase in UBC level. This means that with the combination of low NO_x burners and two-stage combustion, NO_x emission levels <200 mg/m_o^3 can be obtained for a wide range of coals. Only very low volatile and very low reactive coals should be excluded.

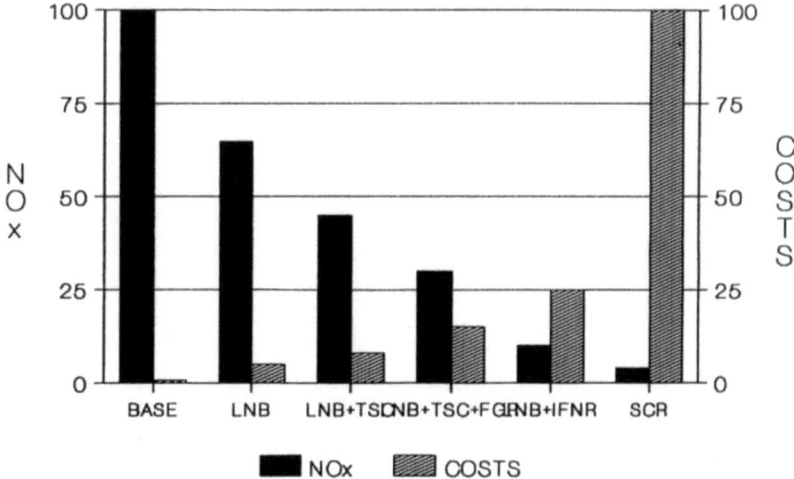

Figure 13: Relative costs and performance of different NO_x emission reduction systems for a retrofitted gas fired unit

HIGH AND LOW DUST SCR PROCESSES

E. Weber and D. Schmidt
Institut für Umweltverfahrenstechnik
Universität Essen
D-4300 Essen, Federal Republic of Germany

1. Introduction

In all countries where secondary methods (applied after the formation of pollutants, i.e. post combustion control in the case of NO_x) are used for the control of NO_x emissions from stationary sources, processes of Selective Catalytic Reduction (SCR) play a significant role. Depending on the state of the gas to be treated by SCR, the NO_x reduction is classified as a high or a low dust process. The NO_x removal is called selective because the reduction chemicals should react exclusively with the pollutant. Ammonia is applied as a suitable reduction medium on a technical scale.

Taking into account that more than 95% of the nitrogen oxides in flue gases from the combustion of fossil fuels consist of NO, the reduction of NO_x can be described by the following overall chemical equations:

$$6\,NO + 4\,NH_3 \longrightarrow 6\,H_2O + 5\,N_2$$

and

$$4\,NO + 4\,NH_3 + O_2 \longrightarrow 6\,H_2O + 4\,N_2$$

Because of the presence of oxygen in the flue gas and in accordance with the catalytic reaction mechanism of most industrial catalysts, the second equation has usually to be favoured.

From the large number of possible reactions that may take place simultaneously, the formation of nitrogen oxide by oxidization of ammonia is of major interest:

$$4\,NH_3 + 5\,O_2 \longrightarrow 4\,NO + 6\,N_2O$$

This reaction, occurring at higher temperatures, reverses the NO_x reduction process.

It is remarkable that the overall equations for both the catalytic and the noncatalytic reduction of NO by ammonia are the same, although the underlying reaction mechanisms are quite different. The application of a catalyst enhances the reaction rate of the NO reduction and thus allows lower operation temperatures and in comparison to non-catalytic processes higher NO conversion rates at lower molar NH_3/NO ratios.

The obtainable high NO conversion rates, the technical and economic feasibility and the fact that the products of the process, nitrogen and water, are natural constituents of the atmosphere are the main causes for the attractivity of SCR processes for secondary NO_x control.

2. Installation of Catalytic Reactors

The utilization of catalysts for the removal of NO_x from flue gases of fossil-fueled boilers on an industrial scale started about twenty years ago in Japan. At the end of this development, the installation of catalysts in the flue gas stream between boiler and air preheater at temperatures between about 300 to 400°C seemed to offer the optimum solution. The reactors had been designed as high dust reactors without previous dust removal or as low dust installations with a hot gas electrostatic precipitator positioned upstream.

As NO_x control aroused growing attention in Europe and in the USA, other possible positions for the installation of catalytic reactors have been discussed and investigated. Figure 1 shows schematically how catalysts for NO_x reduction can be installed in the flue gas stream of a fossil-fueled facility. SCR 1 and SCR 2 demonstrate the high and low dust operation mode of catalytic reduction between boiler and air preheater.

Besides of Japan, where this solution had been chosen first, as already mentioned, there are several installations of this type in European countries and in the USA.

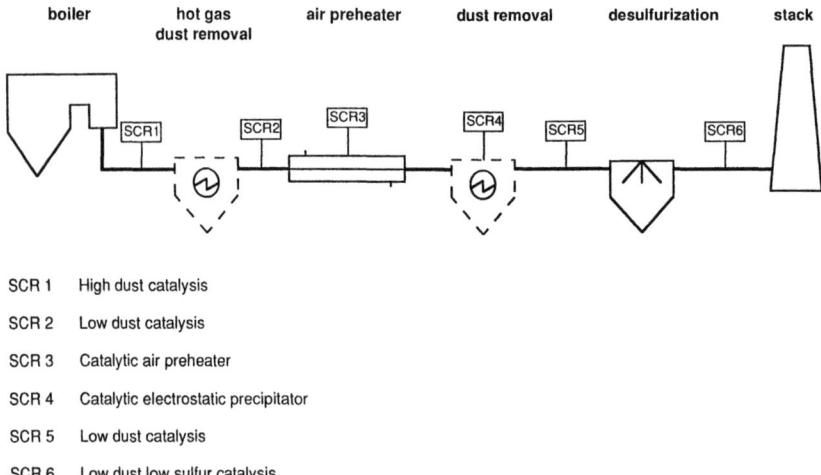

SCR 1   High dust catalysis
SCR 2   Low dust catalysis
SCR 3   Catalytic air preheater
SCR 4   Catalytic electrostatic precipitator
SCR 5   Low dust catalysis
SCR 6   Low dust, low sulfur catalysis

Figure 1. Installation of SCR processes.

Until now, the processes SCR 3 or SCR 4, which use modified air preheaters or electrostatic precipitators as catalytic reactors, have not been applied in industrial plants, although in principle both devices offer the advantage of sparing a separate reactor for the NO_x reduction. For the catalytic electrostatic precipitator, laboratory investigations have proved that the application of an electrical field can enhance the NO conversion rate by a factor of two. Separation efficiencies of about 40% increased up to more than 80%. The installations SCR 5 and SCR 6, at positions where either dust or dust and gaseous pollutants are already removed, have been realized in various power stations, especially in the Federal Republic of Germany. Since SCR 3 and SCR 4 have not been applied on a technical scale, SCR 1 is at present the only available high dust process.

Due to the absence of dust particles in the catalytic reaction zone, low dust processes show the following advantages in comparison to high dust systems:
- No problems of clogging and sticking dust particles in the gas flow channels of the catalyst.
- No abrasion or erosion of the catalyst's surface by the transport of dust.
- Contamination of dust by ammonium or ammonium salts is avoided.
- No poisoning or inactivation of the catalyst by dust constituents.
- Smaller cross sections for the gas flow channels in the catalyst are possible, allowing a larger specific surface area and thus a higher reactivity of the catalyst per reactor volume.

In spite of these items, high dust processes have been preferred in many cases:
- The application of hot gas electrostatic precipitators is not very convenient, especially in Europe. Barrier filters for high temperatures are available only very recently.
- Due to the increase of gas velocity with higher temperatures, larger-sized units are needed for the dust removal at elevated temperatures.
- In many already built facilities there is not enough space available for the post-installation of both, a catalytic reactor and a hot gas clean-up system.

The installation between the boiler and the air preheater (SCR 1 and SCR 2) offers the easy availability of flue gas temperatures which are in the optimum activity range for most catalysts with transition metal compounds as active components.

Besides that, the process temperatures are sufficiently high to avoid deposits of ammonium salts which may be formed by reaction of ammonia with the acidic constituents of the flue gas. Due to their relatively high thermal stability, ammonium salts of sulfuric acid (solid $(NH_4)_2SO_4$ and molten NH_4HSO_4) are the salts to be most probably formed at higher temperatures. However, even the highest SO_3-concentrations to be expected (about 50 ppm) in flue gases rich of oxygen and sulfur dioxide, are not of the order of magnitude for a probable deposition of ammonium salts. On the other hand, it was a great problem in the early days of catalyst development to overcome the SO_2/SO_3 conversion which is catalysed by some of the $DeNO_x$-active materials as well.

The installation of a catalytic process downstream of the air preheater after the dust removal has the same advantages as the low dust process SCR 2. The flue gas temperatures, however, are already so low that ammonium salts of the sulfurous acid are formed which would deposit and inactivate the catalyst's active surface. Hence, it is necessary, either to heat the gas to temperatures above the temperature of deposition or to desulfurize the gas before it enters the catalytic reactor. In the case of ceramic catalysts, usually the flue gas is warmed up by heat exchange to meet the activity optimum of the catalyst. Activated carbon catalysts, on the other hand, allowing reaction temperatures of about 100°C can be operated without heating, if the gas is desulfurized fairly well down to concentrations of a few ppm SO_2. The installation of the SCR process at the end of the gas cleaning steps after dedustion and desulfurization has the advantage that by the several upstream clean-up processes components which might be harmful to the catalysts have been removed to a far extent.

3. Performance and Design of Catalysts

Of the various possible designs and shapes that in principle can be used for SCR, only a few types are actually in operation in industrial facilities. Figure 2 surveys some important types of these catalysts. Whereas granular beds are applicable only for lower dust concentrations, plates and honeycomb catalysts can be used also for high dust systems.

In the case of high dust SCR, the catalysts are usually installed with wider, vertically orientated gas flow channels to avoid clogging effects by the dust.

Honeycomb catalysts are preferentially used in form of modules with a rectangular design of the gas flow channels. The honeycomb structure is formed by a more or less homogeneous porous ceramic material, doped with catalytically active components.

Plate-shaped catalysts may consist either of porous ceramic material alone or more often of ceramic layers which are supported by sheets of metal. Other types like tube-sized or honeycomb metal catalysts are not used on an industrial scale for the clean-up of flue gases.

The thickness of the catalytic active porous layers in ceramic honeycomb or plate-shaped catalysts is varying between about one and a few millimetres. The minimum distances of the catalytic layers, i.e. the distance of the plates or the minimum width of the gas flow channels, is ranging from about 1 mm (for low dust SCR) to approximately 10 mm (for high dust application).

Granular or pellet-sized catalysts for the application in low dust installations consist either of activated carbons or of ceramic materials, like those used for honeycomb and plate-shaped catalysts.

Besides a few applications of zeolithes, iron oxide or platinum doped catalysts the majority of NO reduction processes for flue gases is operating with titania-based materials containing preferentially vanadium and tungsten compounds as active components.

Instead of titania, catalysts with a support of alumina or aluminium silicate can only be used in desulfurized or sulfur-free flue gases, because otherwise the supporting materials will be destroyed by the chemical attack of SO_3 sulfuric acid.

4. Reaction Mechanism and Activity of Catalysts

In the presence of oxygen, the NO reduction by ammonia over ceramic catalysts doped with transition metal compounds is in many cases in accordance with an Eley-Rideal mechanism: Ammonia is chemisorbed on acidic surface sites of the catalyst and reacts

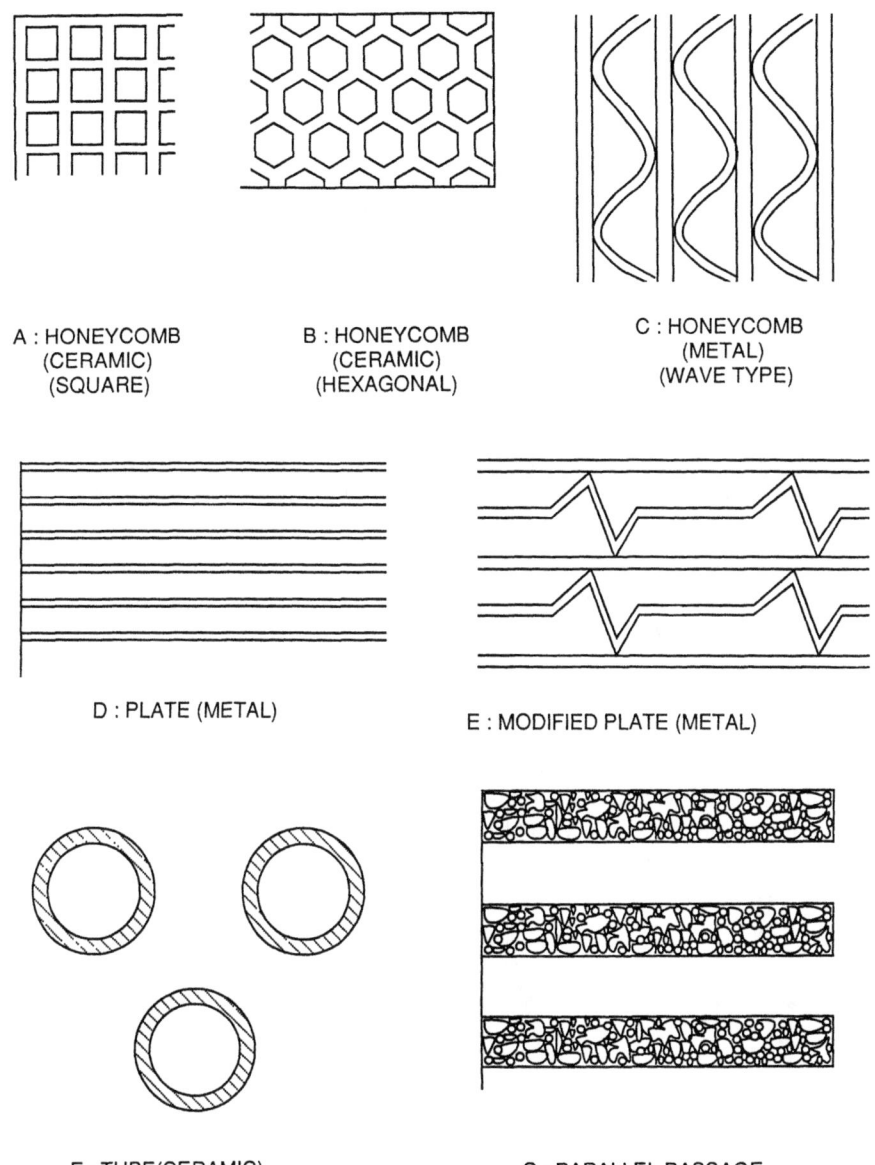

Figure 2. Various shapes of catalysts.

with nitrogen oxide from the gas phase. The order of the reaction rate is zero with respect to ammonia and one for NO, thus the resulting overall reaction may be described by a pseudo-first order equation. Due to this reaction mechanism, the NO conversion is independent of the initial NO concentration. Considering the catalytic reactor as an ideal flow tube, the NO reduction for a given type of catalyst may be written as:

$$\ln(1 - \eta) = -k/SV$$

with:
η = the degree of NO conversion;
k = the rate constant;
SV = the space velocity;
ln = the natural logarithm.

The rate constant k is related with the specific properties of a selected catalyst and usually shows an exponential temperature dependence according to the Arrhenius-equation. On the other hand, it has to be taken into account that with increasing temperatures the oxidation of NH_3 into NO may be enhanced as well. Figure 3 demonstrates the influence of the reaction temperature on NO reduction for the example of titania catalysts with vanadium as active component. The results have been obtained from laboratory experiments studying the NO conversion in synthetic gases with a NO concentration of 500 ppm NO and an oxygen content of 5%. Besides these constituents the test gas consisted of nitrogen as an inert carrier component.

As can be seen, the degree of NO conversion increases at first with growing temperatures, starting at about 150°C, then reaching maximum values of about 100% near temperatures of 250°C. Around 400°C the NO conversion is decreasing, due to the onset of the starting ammonia oxidization.

A similar behaviour has been found for various types of $DeNO_x$ catalysts and even for the noncatalytic $DeNO_x$ process. The width of the active temperature window for the NO conversion, the temperatures at which it starts and the achievable NO conversion rates differ. In the case of noncatalytic reactions, the reduction starts at temperatures near 800°C, the width of the window is about 100°C and the conversion rate is much below 100%, even for molar NH_3/NO ratios above one. Hence, it can be stated that the application of a catalyst successfully lowers the activation energy for the NO reduction and allows more convenient reaction temperatures.

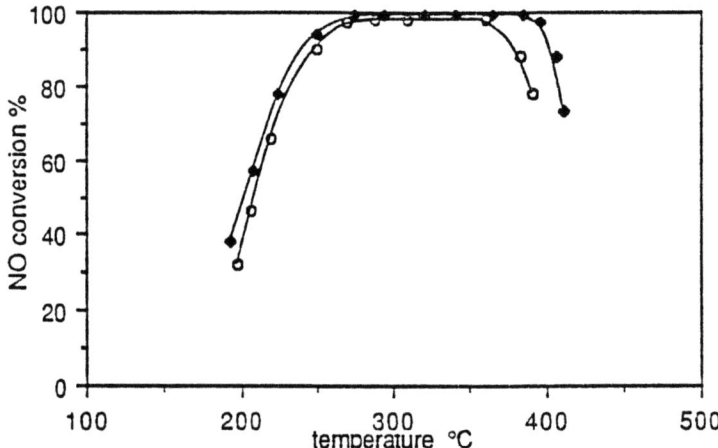

Figure 3. Temperature dependence of NO conversion for titania-based catalysts.

In principle, the rate equation may be modified by introducing a surface specific rate constant k', due to the aspect that the rate of heterogeneous catalytic reaction is proportional to the surface area A:

$$\ln(1 - \eta) = k' \, A/SV$$

The space velocity SV is given by the ratio of the volume flow rate to the catalyst volume and is thus proportional to the reciprocal of the mean residence time of the gas in the catalyst.

Comparing values of the space velocity from literature date, it is necessary to verify whether the gas flow was calculated in the actual or normal state and whether the catalyst's volume was thought of as the total reactor volume or only as the free space of the gas channels. The surface area may be considered as the outer geometric surface area or as the gas-admittable inner surface area which can, for instance, be obtained by gas adsorption experiments.

The gas-admittable surface of various catalysts usually ranges from about 30 to more than 100 m^2/g of catalytic material to get a sufficiently high reactivity.

The constant k of industrially applied catalysts belonging to the type that reacts according an Eley-Rideal mechanism is about 7500 h^{-1} at optimum reaction temperatures (above 300°C), calculated in the normal state and for a catalyst-free reactor volume.

For a catalyst running at a space velocity of SV = 3000 h^{-1}, a degree of NO reduction of about 91,8% is calculated from the rate equation. Assuming that a coal-fired power station delivers 3000 m^3 in the normal state per hour and installed electric megawatt of electric power, it appears that a catalyst volume of 1 m^3 is needed per megawatt of electric power.

Other catalysts, like zeolithes, have an activity that is about two thirds of vanadium-doped titania catalysts.

Nonceramic catalysts (e.g. activated carbons and cokes) and ceramic catalysts with noble metals as active components react according a Langmuir-Hinshelwood mechanism: NO and NH_3 react in the chemisorbed state. In the case of carbons, it is not a solely catalytic reaction. A small amount of NO, the concentration of which is depending on the reaction conditions, is directly reduced by carbon to N_2. A disadvantage of the use of noble metals (e.g. platinum) as active components, besides the relatively high costs, may be that the catalysts form significant concentrations of N_2O instead of N_2 as a product of the NO reduction.

In agreement with the reaction mechanisms a poisoning of the catalysts may occur by chemical reaction with other components of the flue gases. Alkaline materials can neutralize the acidic surface sites and thus inhibit the chemisorption of ammonia. Condensable components like ammonium salts, arsenic and volatile silicon or iron compounds may inactivate the active surface sites of catalysts by forming deposits which can not be permeated by the gas components. It is therefore necessary to choose well the proper operation conditions for SCR processes in order to achieve sufficiently long life expectations.

5. Ammonia Clean Gas Concentration

Unaffected by the kind of process, the technical application of SCR is always connected with the occurrence of ammonia in the cleaned gas leaving the catalytic reactor. At lower temperatures, ammonia will react with acidic gas constituents in forming ammonium salts. Figure 4 surveys formation and decomposition temperatures for various ammonium salts. As already mentioned, the most important components are the salts of sulfuric acid.

These salts may cause problems by:
- Emissions into the atmosphere (blue haze).
- Deposits in heat exchangers and other components of the plant along the flue gas line.
- Depositions onto the fly ash.
- Enrichment of ammonium in the flue gas desulfurization system.

Investigations on the high dust process SCR 1 have shown that, regardless of the conditions of operation and the ammonia concentrations, more than 90% of the compounds are found in the air preheater and the fly ash. The concentrations of ammonia salts in the air preheater varied between about 30 to 50%. The higher ammonium salt concentrations in the air preheater have been observed at lower gas inlet temperatures. In most cases the largest amount of ammonia salts was carried over to the fly ash. A maximum value of about 5% of the total concentration of NH_3 leaving the catalytic reactor was passing into the flue gas desulfurization system and less than 1% was found to enter finally the atmosphere.

The enrichment of ammonia salts in the fly ash may be problematic for the further use of the particulates in the cementum industries or for pavement purposes. The smell of ammonia is, at least for psychological reasons, not acceptable in fly ashes to be used as

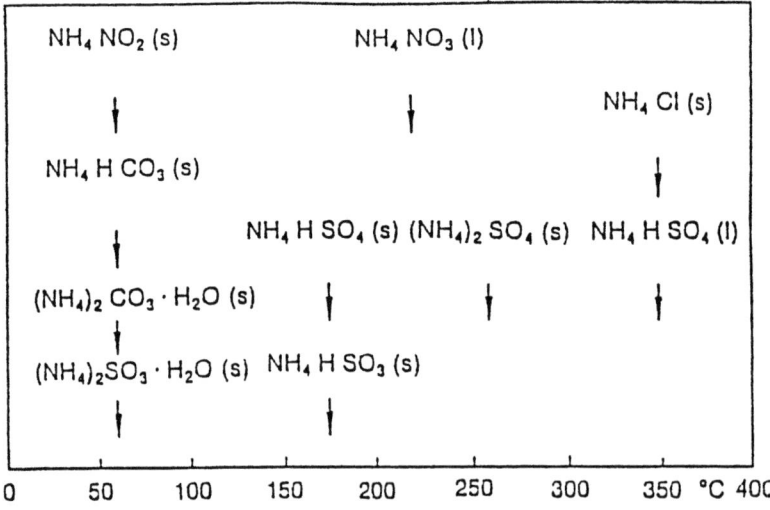

Figure 4. Formation and decomposition temperatures of ammonium salts.

a construction material. Therefore, it might become necessary to remove the ammonium compounds or the ammonia from the dust sampled in the hoppers of the electrostatic precipitator. Problems may occur especially when the clean gas concentration of NH_3 increases from a value of about less than 1 ppm for newly installed catalysts to more than 5 ppm after longer operation periods or if operation failures occur.

To remove the ammonium salts from the fly ash, it will be necessary to heat the dust to temperatures of more than 300°C. Processes have been proposed to reduce the ammonia contents in a fluidized bed reactor by means of superheated steam with addition of a small amount of quick lime.

Low dust processes may have similar problems with heat exchanger surfaces, but not with ammonia loaden dust because the dust does not come into contact with ammonia or ammonium salts. In general, heat exchangers will have to be cleaned from time to time.

For installations of low dust processes behind dust precipitation and desulfurization at the "cold end" of the flue gas line, it must be certified that emissions of ammonium salts to the atmosphere can be controlled under all circumstances of operation. This demands for larger dimensions of the catalytic reactor and a sufficiently understoichiometric application of ammonia. On the other hand, due to the operation in a clean gas atmosphere, the life expectations for catalysts may be comparatively enlarged.

6. Realization and Performance of SCR Processes

Constructions and conceptions of SCR processes for secondary NO_x control in Europe started relying on the experiences in Japan and on own investigations. It appeared that high dust SCR processes successfully approved in Japanese power stations over several years, could be transferred without major modifications to European installations for applications behind coal-fired burners with dry ash removal. The use of SCR systems for flue gases with low dust and sulfur oxide concentrations obviously could be realized without difficulties, too.

On the other hand, investigations carried out at several pilot plants showed that problems occurred with slagging ash boilers, especially with installations with a feed-back of the fly ash to the boiler. In these systems, increasing contents of arsenic have been observed which finally caused a relatively quick inactivation of the catalysts by deposits nonpermeable to the gaseous constituents. Thus, the life expectations of catalysts decreased to less than 1000 hours. Apparently, it became necessary for this type of coal-fired boilers to install the SCR system at the cold end of the flue gas line where the arsenic already had been removed from the flue gas by previous condensation on dust particles.

For instance, in the Federal Republic of Germany at the end of a fast development, accelerated by legislative demands, more than 50% of the secondary NO_x control systems are of the high dust SCR type and more than 30% are of the low dust SCR type are installed behind the flue gas desulfurization process.

A comparison between low dust SCR processes installed between the economiser and the air preheater and those behind the desulfurization system shows the following main differences:
- The latter installation needs preheating of the flue gas to arrive at the reaction temperature of the catalyst, which may be somewhat lower than for installations between the economiser and the airpreheater (at least 200 to 300°C for ceramic catalysts, about 100°C for activated carbons). The preheating is usually carried out by regenerative heat exchange as demonstrated by Figure 5. To reach the operating temperature at the start of the catalytic reduction, an oil burner or steam may be used. Figure 5 shows systems with one and two regenerative heat exchangers.
- In the absence of sulfur oxides, catalysts may be applicated that cannot withstand the chemical attack of sulfur compounds.
- Due to the installation at the end of the flue gas line, the SCR process is practically independent of the operation conditions, e.g. flue gas streams of different boilers may be cleaned by one SCR process.

Independent of the type of SCR system, the injection of ammonia into the flue gas has to be carried out in such a way that the ammonia is well-dispersed in the flue gas to avoid large slips of ammonia to the clean gas side and to achieve the maximum possible degree of NO conversion.

The ammonia can be supplied as liquefied ammonia of a technical degree of purity or as an aqueous solution containing about 25% of ammonia. Due to its lower cost, the liquid ammonia is preferentially used as a reducing medium for larger flue gas streams whereas the aqueous solution is favoured for smaller installations, taking into account

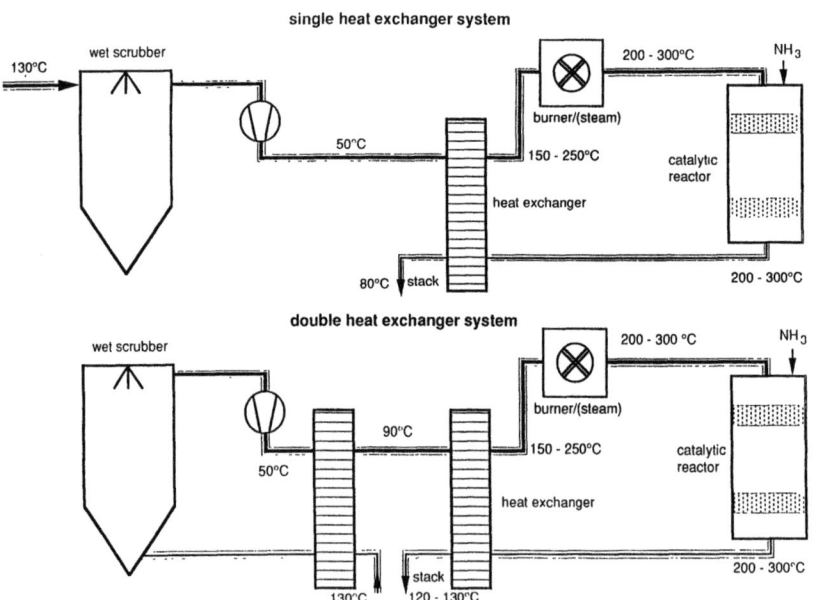

Figure 5. Installation of low SCR dust after wet scrubber (SCR 6).

that security aspects cause increasing specific costs for smaller plants. Another point of view to prefer the choice of the aqueous ammonia solution is that security demands may in some cases not allow the transport of liquid ammonia to the plant. Particularly, it must be taken into consideration that liquid ammonia has to be stored and transported either at low temperatures (below -35°C) or at elevated pressure. Aqueous solutions may be handled at ambient temperatures and ambient pressure.

For smaller installations, aqueous NH_3 can be fed directly to the flue gas stream in most cases of application. However the aqueous solution is normally injected by means of a transport gas medium such as hot air or steam. From an energetic point of view, it is an advantage to use fractionated distillation or rectification to get higher concentration of ammonia in the gas phase. Similarly, liquid NH_3 may be injected with the aid of cold air into the flue gas.

Uninfluenced by the used species of the reduction medium, the application of the transport medium makes it necessary to have a mixing chamber in which the ammonia is well dispersed.

The transport medium/ammonia mixture is then finally supplied to the flue gas by a venturi tube, an evaporator or preferentially by means of a nozzle system to certify that the whole cross section of the flue gas channel shows a constant concentration of ammonia in front of the catalytic reactor.

The amount of ammonia to be injected into the flue gas stream is usually controlled by the flow rate of the flue gases and the inlet gas NO concentration. To avoid significant clean gas concentrations the molar NH_3/NO ratio is usually chosen smaller than 1 (at about 0,9 to 0,95).

The linear gas velocities in the flow channels of a catalytic reactor may be chosen between about 4 m/s to 10 m/s. It must be taken into account that the choice of a suitable gas velocity is in particular critical for high dust SCR processes: Too low values, (less than about 6 m/s) will usually result in clogging effects, although the flow channels are vertically orientated for high dust installations. On the other hand, values of more than 8 m/s will not only increase the pressure drop but also will enhance the corrosion to intolerable levels. A small erosion effect may have some advantages for high dust SCR by abrasing deposits on inactive surface sites.

The pressure drop of an industrial ceramic catalyst which has been designed to remove more than 90% of No_x is in the range of about 500 to 900 Pa.

For the design of a low dust SCR process installed behind the flue gas desulfurization system, the demands for low ammonia emissions to the atmosphere make it necessary to select a larger active surface in comparison to a high dust process to avoid high ammonia emissions. These are overcome otherwise by the uptake of the ammonia by the fly ash and by the desulfurization system.

In can be summarized that for new installations in coal-fired boilers with solid ash removal, high dust processes are usually the most acceptable way of secondary NO_x control. In other cases it depends on the circumstances which type of low dust SCR shall be chosen.

7. References

Haagen Smit, and Fox, M.M., "The chemistry of photochemical smog", Ind. Eng. Chem. 1956, 48, 1484.

Ando, J., "SO und NO_x Removal for Coal-Fired Boilers in Japan", Seventh Symposium on Flue Gas Desulfurization Hollywood, Florida 17.-20.05.1982.

Ando, J, Stern, R.D. and Mobley, J.D., "Status of Flue Gas Treatment Technology for Control of NO_x Simultaneous Control of SO_x and NO_x in the United States and Japan", AlChe (Proc.) 69th Ann. Meet. Chikago 1976.

Wahl, D.J., "Neueste Erfahrungen mit den Entstickungsanlagen der VKR", VGB Kraftwerkstechnik 67 (1987), Nr. 12, S. 1198-1203.

Erath, R., "Einsatz der SCR-Technologie im High-Dust-Bereich bei Schmelzkammerfeuerungen unter besonderer Berücksichtigung der Feuerungen mit Ascherückführung", Technische Mitteilungen 80 (1987), Nr. 9, S. 529-596.

Schallert, B., "Erfahrungen aus einem zweijährigen Versuchsbetrieb mit DENOX-Pilotanlagen", VGB Kraftwerkstechnik 68 (1988), Nr. 7, S. 432-440.

Haji Iavad, M., "SCR-Anlage vor LUVO oder nach REA", BWK, Bd. 37 (1985), Nr. 1-2, S. 39.

Schallert, B. and Kaulitz, I., "Betriebserfahrungen mit SCR-DENOX-Versuchsanlagen unter verschiedenen Einsatzbedingungen", VGB Kraftwerkstechnik 66, Heft 9, Sept. 1986.

COSTS OF DESULPHURISATION AND DENOXING

D. VAN VELZEN
Commission of the European Communities
Joint Research Centre,
Environment Institute
I-21020 ISPRA (Varese) Italy

1. Introduction

Measures for the abatement of environmental pollutants are now generally accepted as being necessary. They are applied on a very large scale, resulting in a considerable expenditure. In fact, it is an unprecedented phenomenon in the story of industrial development that investments of this magnitude have been carried out by industry for purposes not directly aiming at profit.
In Western Germany there are now 165 flue gas desulphurisation plants in operation in 72 sites of public power stations. This is practically the totality of the coal fired power stations (28 GW_{el} coal and 11 GW_{el} lignite). According to an inquiry of the association of electricity producers (VGB), the investments done for this purpose amount to 14.2 milliards DM = 6.9 milliards ECU [1]. It is expected that for the reduction of the NO_x emissions from power stations, additionally investments of 7 milliards DM are required. These investments are foreseen for the installation of 33 GW_{el} flue gas denoxing plants and for primary measures during combustion. It follows that flue gas desulphurisation and denoxing is expensive.

From the above figures it can be calculated that the average investment for a flue gas desulphurisation process is about 180 ECU/kW_{el}. The deviations from this average value for individual plants are very large, as the real costs depend on the sulphur content of the coal, local conditions, the question if the installation is a new one or a retrofit etc. etc. This illustrates the difficulty encountered by the reader who is interested in a fair cost comparison between one or more concurrent desulphurisation or denoxing processes. Results from cost calculations in the open literature are often quite different, even for the same process.

To give an example, in a recent invited survey lecture [2], the renowned Japanese expert Ando stated that the actual investment costs for the Selective Catalytic Reduction (SCR) denoxing process is 50 - 70 $/kW (40 - 57 ECU/kW) and that the removal of 1 ton of NO_x may cost $1000 - 1500 (810 - 1220 ECU).
At nearly the same time Schärer and Haug [3] gave an extensive survey for the German situation [3]. They mentioned a figure for the capital requirements for the SCR process in the order of 30 DM/kW_{th}, which is equivalent to an investment of approximately 40 ECU/kW_e. The separation costs for NO_x are quoted as 2 DM/ton, i.e. 970 ECU/kg.
The problem was also treated from the point of view of a U.S. based catalyst manufacturer by Boer et. al. [4], who differentiated between first and second generation catalysts. Their most favourable example (second generation catalyst

and low NO content in the flue gas) gave values for the capital costs of 74 $/kW and NO_x removal costs of 400 $/ton.

A comparison of the above data leads to the conclusion that the Japanese and German data are more or less comparable, the German data being somewhat lower. The most striking difference, however, is between the U.S. and the other two figures. Particularly surprising is the observation that the U.S. source gives the highest value for the investment costs (60 ECU/kWe against a range of 40 - 57 ECU/kWe), whereas at the same time the quoted total process costs are only about 40% of the Japanese and European figures. Given that all authors are reporting about the same chemical process, one must conclude that there are fundamental differences in the applied cost calculation procedures and that, thus, the results are not comparable. None of the authors give indications about their cost calculation techniques, so that it is virtually impossible to find out which is the cause of the (macroscopic!) differences.

This is indeed the main obstacle in the correct judging of published cost data. Reported costs are only comparable when the same cost estimation techniques are applied, otherwise the results reflect mainly the basic differences of the various working hypotheses. It is therefore necessary to explain in detail the cost analysis technique applied in the present study. This will be done in the next chapter.

Finally it has to be noted that the reader must be particularly attentive and critical in cases where newly developed processes are announced by scientists who developed the process on a laboratory or bench scale. These publications often contain chapters about the economic outlooks of the process in question, frequently containing phrases like "considerably more economic than current processes" or statements of the same nature. These statements are nearly always based on very approximate cost calculations, neglecting cost factors like engineering costs, overheads, process contingency, maintenance etc.

2. Cost estimating procedures

2.1 General

Several methods of varying degrees of accuracy are available for estimating the capital and operating costs of chemical processes. The methods range from very approximate estimations using only basic process data to very detailed cost estimates based on preliminary design, contractor quotes and process lay-outs. Examples of methods of the low end of the range are the Stallworthy method [5] and the Functional Units Method [6]. In the last method the expected investment costs are simply calculated by the formula:

$$I = F \times N \times CF \times (CP \times R^{0.333})^{0.66}, \text{ where}$$

I   = Investment costs
F   = a constant
N   = number of main process units
CF = complexity factor 10^y, y = FM + FP + FT
FM = material factor (0.0 - 0.4)
FP = pressure factor (0.1 log p (bar))

FT = temperature factor (0.018 per 100°C)
CP = plant capacity (ton/h)
R = ratio between the average stream and plant capacity, both in mass units.

The method is based on average costs of many different installations and various chemical processes. However, the degree of accuracy produced using such estimating methods will, at best, be in the "order of magnitude" category.

The detailed cost estimate, in turn, can produce accuracies of about 5 - 10% depending on the amount of preliminary engineering involved. These estimates take many months of engineering effort and are thus very expensive. Detailed process and engineering flow sheets, material and energy balances, instrumentation plans and equipment arrangement drawings have to be prepared before a final cost estimate can be developed.

For comparative cost studies, cost estimation techniques must be applie which are a compromise between the two extremes and produce accuracies of approximately 20%. Standardised methods have to be used for this purpose. The extensive and valuable work done by the Electric Power Research Institute (EPRI) in Palo Alto (California - USA) can form the basis for a valuable general model. EPRI carried out an impressive number of comparative FGD cost studies [7,8] all based on application in a hypothetical new high sulphur coal power plant of 1000 MW (two 500 MW units). A flow sheet, material balance, equipment list and utility consumption form the basis of each evaluation. Equipment cost information was obtained from process vendors and capital costs were estimated by factoring cost items as a function of equipment costs. Operating costs were estimated from reagent and utility consumption. Overall results were reported to have an absolute accuracy of 30%, whereas the relative accuracy for process comparison purposes is estimated at 15%.

The EPRI conventions differ somewhat from standard practice in modern European process plant industry (process redundancies, many multiple trains...). In a study, overviewing a large part of FGD cost data originating from USA, Japanese and European sources, IEA Coal Research [9] applied a slightly modified and simplified form of the EPRI convention, which was also used in a recent study by Pax Technology Transfer for the European Comisssion [10]. In the present paper we will apply this convention.

2.2 Capital Costs
Capital costs are broken down in the following categories:
- Total Plant Costs (TPC)
- General Facilities
- Engineering
- Royalties
- Contingencies
- Start-up Costs
- Working Capital
- Interest during Construction (IDC)
- Land Costs.

The sum of the total gives the Total Present Value (TPV) of the project, which is then the basis for the calculation of the Direct Cash Flow which is one of the main factors in the calculation of the running costs of the process.

2.3 Capital Cost Factors
The convention applied in the present study consists of the detailed calculation of the total plant costs (TPC). This is the sum of the costs of all equipment and non-consumable materials within the battery limits of the FGD plant. Like in the EPRI methodology, the determination of the TPC is based on process flow sheets, material balances, equipment specification and cost information from equipment suppliers. It includes piping, instrumentation, process control equipment, civil work and site preparation. The purchase of the land required by the FGD plant battery limits has also to be included in the TPC.
The values of the other capital cost items (general facilities, etc.) are not individually calculated, but estimated as fractions of the TPC by the application of so-called cost factors. The numerical values of the cost factors are based on statistical experience of similar types of process plants. The various cost items to be determined by the cost factor approach are discussed in detail below.

2.3.1 General Facilities This is the total costs of permanent non-plant infrastructural facilities required for the FGD or denoxing project, including a share of the common facilities. They include roads, buildings, laboratories, workshops etc. Raw material and products storage facilities do not belong to this category, they have to be included in the TPC. The cost factor applied for the general facilities costs is ***0.1 x TPC***.

2.3.2 Engineering This is the cost for process design and management throughout the total process construction phase. It includes the direct and overhead contractor costs, including the contractor's profit. The process industry term "engineering" is generally employed for this cost item. Its cost factor amounts to ***0.125 x TPC***.

2.3.3 Royalties Most FGD and denoxing processes are proprietary licensed technologies for which the user is liable to pay royalties to the patent owners. The height of the sum for royalties is varying widely and it is not easy to get reliable information about this cost item. The present and the cited cost studies use, as an average, a cost factor equal to ***0.01 x TPC***.

2.3.4 Project contingency The project contingency is an additional capital budget item intended to cover additional plant items, modifications and other capital costs not foreseen in the initial budget. They usually result when a completely detailed design has to be carried out on basic plant specifications. It is effectively a measure of the quality of the initial budget estimate. This item is comparable to the "design contingency" concept applied in the process industry. Its cost factor is a sliding one from **0.15 to 0.30 x TPC**. Higher contingency factors are applied to equipment items of special design, the lower end of the scale applies for standard equipment items. In most cases the applied factor for FGD processes is **0.21 x TPC**.

2.3.5 Process contingency is another capital cost factor related to the extent in which the applied technology is proven and demonstrated. There is always a risk in the application of a new technology, which has to be expressed by a risk factor like the one in question. The process contingency is thus related to the level of process development by a scale **between 0.10** for industrially demonstrated process to **0.50 x TPC** for processes developed only at bench scale. For processes backed by the operation of a pilot plant the process contingency factor is between 0.20 and 0.35 TPC.

2.3.6 Total Fixed Investment (TFI) The sum of the TPC, general facilities, engineering and contingencies is called the Total Fixed Investment (TFI). Some costs are considered to be dependent on TFI rather than on TPC.

2.3.7 Start-up or preproduction costs This item is designed to cover the usually encountered start-up and commissioning costs such as operator training and unforeseen events during the start-up of the plant. It is considered to be equal to **one month of operating costs plus 0.02 x TFI**.

2.3.8 Working capital This cost item is considered the capital required to provide **one month of the variable operating costs**.

2.3.9 Interest during construction (IDC) These are the costs of interests on the funds during the construction phase, where the operating cost calculations are not yet applicable. The value of the item is drawn from a postulated capital drawdown schedule and is usually estimated as **0.1 x TFI** .

2.4 Operating costs
Operating costs are subdivided into two categories: variable and fixed.

2.4.1 Variable operating costs are those which vary directly to plant production rate. They consist primarily of the cost of raw materials, utilities and other process consumables, such as chemicals. In the case that saleable by-products are produced, the credits obtained are put in diminuition of the variable operating costs. In the present study, the variable operating costs are calculated on an hourly basis at the maximum specified plant capacity. Annual variable operating costs are derived from the hourly costs by assuming a load factor of 70%, i.e. the plant is operating 6132 hours per annum.

2.4.2 Fixed operating costs are those which are independent of the production rate. The principal components are direct and indirect labour and overhead expenses. In the present convention, also maintenance costs are considered under fixed operating costs. The EPRI convention proposes the application of a cost factor between 0.015 and 0.10 x TPC per year, depending upon the plant conditions. For simplicity, IEA and Pax applied a single factor, 0.05 x TPC, for all FGD processes. Maintenance costs can be subdivided into material and labour costs. Generally a 60/40 split between materials and labour is applied. However, this subdivision has no influence on the total maintenance costs in the present convention. The components of the fixed operating costs are summarised as follows:
1) Direct labour costs: number of applicable personnel x average labour costs x 8760 hours per year. These costs have to include a payroll burden. In this study it is postulated that the direct operating labour rate is 17.50 ECU/h. The payroll burden is assumed to be 50%, so that the labour costs amount to 26.25 ECU/manhour.

2) Overall maintenance costs: 0.05 x TPC/y
3) Indirect labour costs for administration and support services: 30% of the sum of direct labour and maintenance labour costs.
4) Taxes and insurance: 0.032 x TFI/y.

2.5 Levelised and total annual costs

It is consistent with the present convention to adopt a simplified approach to take account for the relationship between cash flow and time. The present study adopts the project financial parameters employed in [9] in order to derive Direct Cash Flow (DCF) values to cover FGD costs levelised over the project life and for the DCF costs per kWh. The project life is assumed to be 15 years, the discount rate is taken as 5%/y, and no correction for inflation is made.
For these conditions, the levelised annual capital charge factor is 0.0963 x TPV (Total Present Value - Sect. 2.2).

The total annual costs are then calculated as the sum of
- fixed operating costs
- variable operating costs
- levelised annual capital charge.

The annual cost/charge per net kWh produced at the station busbars is calculated as the ratio of the total annual costs and the net energy produced, i.e. in 6132 hours of operation at 100% capacity, taking into account the energy consumed by the FGD process itself.

3. Plant module concept

Reliable published cost data for FGD processes are generally point values which are the results of a cost study for one specific plant of a given size, a given flue gas flow rate and a given sulphur dioxide content. It is desirable to be able to use these data as a basis for the extrapolation to other conditions of flue gas and sulphur dioxide flow rate. This is the only way to perform a comparative cost study for various FGD processes for a large range of plant capacities and using different fuels with different sulphur contents. In the Pax study a analytical methodology for comparative cost evaluation was proposed, based on the differentiation of the FGD process in plant modules [10].

Any FGD process contains a group of components which are primarily determined by the flue gas flow rate (columns, ducts, fans...) and a second group which is closely related to the quantity of sulphur dioxide processed. These two groups of components are physically distinct. The capital as well as the operating costs for each of the groups can be separately identified.

The capital costs for the flue gas (FG) module are a function of the design maximum flue gas flow rate. Variable operating costs are proportional to the actual flow rate. Its main component is the energy cost for the transport of the large amounts of flue gas.
The capital costs for the sulphur (S) module are determined by the maximum amount of sulphur removed, whereas the variable operating costs are largely

determined by the reagent consumption and the costs of the product preparation.
There is, of course, also a third group of facilities, which is common to both other groups, e.g. control rooms, utilities etc. Normally, the value of this last group is relatively small compared with the total plant costs. For this reason, these general costs are grouped within the flue gas dependent module.

In the module methodology, published capital cost data are used to establish Base Case Capital Cost Modules, which are employed for estimating the capital costs of the Module for other design capacity conditions using the semi-empirical "two-thirds power rule" which has the form:
$$I = F \cdot (CP^{0.667})$$
This rule is widely used for rapid cost estimation in the process plant industry and is regarded as giving acceptable results if the range over which the extrapolation is done remains between 0.5 and 1.5 times the Base Case Value.
Variable operating costs are directly proportional to the material quantities processed. In the plant module methodology, this proportionality is determined for the FG-module as well as for the S-module, based on the example of the Base Case Module Process. The variable operating costs are expressed as a FG unit rate per million Nm^3 of flue gas and as a S unit rate per kg of sulphur removed. Operating cost modules can then be simply calculated by multiplying the unit rates by the FG and the S flow rate for each specific case.
In certain published cost studies a detailed breakdown for the costs into FG and S modules is not provided. In such cases, the available data are carefully analysed for any inferential basis for the split, including an engineering judgement taking into account of similarities between various published process data.
An example of the applied methodology is given below.

3.1 Limestone/ gypsum (KRC) process
It is required to estimate the effects on the FGD process costs of plant size and sulphur content of a coal fired power station. The cases of interest are capacities of 100 and 500 MWe and 1.0, 2.0, 4.0 and 6.0 wt% of sulphur in the coal.
This example has been treated in the Pax study. As the Base Case Process was chosen a detailed publication of Necker and Strauss [11] about the operation experience at the Neckarwerke. The process in question is the Knauff-Research-Cottrell process (KRC-process), based on the washing of the flue gas with a limestone slurry and production of cement grade gypsum. Process details will not be discussed in this paper, the interested reader is referred to the original study and/or to the presentations on this subject during the present course.

The Base Case Process is a 423 MW coal fired facility at the Altbach/Deizisau power station (unit 5), where the FGD unit is a part of a new integrated power plant and is realised in two phases. It was designed for the use of hard coal with 1.3% of sulphur and the flue gas flow rate at 100% load is specified as 1 490 000 Nm^3/h.
Phase 1 of the project comprised one 100% handling/scrubbing train and a single 100% limestone/gypsum train with associated waste water treatment.
Phase 2 comprised another 100% FG handling/scrubbing train with regenerative reheat of the treated flue gases. The authors state that the Phase 2 scrubber was added to ensure compliance with possible future more stringent sulphur removal

requirements. They also state that one train should be sufficient for the treatment of the specified flue gas flow rate.
The final gypsum processing train including drying, compaction and buffer storage seems to be largely oversized. It was designed for 1 shift in 3 operations.

3.1.1 Capital Costs The reported investments were as follows:
Phase 1:   DM   85 000 000   (1985 basis)
Phase 2:   DM   42 000 000   (1986 basis)
Total: DM 127 000 000
The capital cost modules are estimated from the reported figures as follows:
The reported plant costs are as-built, this means that cost items like engineering, process and project contingencies etc. are already included in the cost figures. The reported cost figures can therefore be considered to be equivalent to the TFI costs of the study. Conversion of the 85/86 costs to a 1988 basis and conversion to ECU yields: Total Investment: 127 000 000 x 1.05/2.06 = ECU 64 730 000.
In the Pax study, on the basis of an engineering judgment, it was proposed to use a 60/40 split for the FG and the S-module.

The FG module calls for 60% of the total costs. It is, according the report, two times oversized, so that the real FG module capital costs can be calculated by:
   $0.6 \times (0.5)^{0.667} \times 64\,730\,000$ =   ECU   24 460 000.
The factor F_{FG} in the formula $I_{FG} = F_{FG} \times CP^{0.6667}$ can now easily be calculated.
For convenience we express the costs in millions of ECU and the capacity in millions of Nm³/h. It follows that **F_{FG} is equal to 18.75**.

The S related portion of the investment costs is
   $0.4 \times 64\,730\,000$ =   ECU 25 900 000.
About one third of this module is attributable to final gypsum processing, which in its turn is 3 times oversized. Correction for this fact is done as follows:
   $(1 - 0.333/3^{0.6667}) \times 25\,900\,000$ = ECU 21 800 000
The maximum fuel input is 140 000 kg/h at a S-content of 1.3% and the specified removal rate is 85%. It follows that the design S-removal rate is equal to:
   $0.85 \times 0.013 \times 140\,000$ = 1547 kg/h S
Expressing the investment costs in millions of ECU and the S removal rate in kg/h, it follows that **F_S is equal to 0.163**.
The obtained values of F_{FG} and F_S allow an easy calculation of the estimated investment costs for the other process conditions.

3.1.2 Variable operating costs The consumable operating materials and utilities required for the process are given in reference [11] for a period when the flue gas flow rate was 1 490 000 Nm³/h and the average sulphur dioxide content of the flue gas was 1700 mg/m³:
   Limestone: 4000 kg/h at a unit price of 20 ECU/ton
   Electric power: 6.6 MW
   Water Treatment costs: 95 DM/h.

The variable costs for the FG module consist only of the power consumption. An arbitrary split of 50:50 is made over both modules. The FG module uses 3300 kW per 1.40 million Nm³ of flue gas, at a unit price of 0.0725 ECU/kWh. The **variable cost FG- module cost factor** is thus:
   $3300 \times 0.0725/1.49$ =   **161 ECU/ million Nm³** of flue gas.

The variable cost factor of the S-module is calculated as follows:
  Limestone: 165 DM/h
  Water treatment: 95 DM/h
  Power: 493 DM/h (3300 x 0.0725 x 2.06)
  TOTAL: 753 DM/h = 365 ECU/h

In the given example the S removal rate is lower than the design value. When the degree of desulphurisation is taken as 85%, the S-removal rate is found by:
  1.40 x 1700 x 0.85 x 32/64   = 1012 kg S/h;
The **variable cost factor of the S-module** becomes thus: **0.361 ECU/kg S**.

3.1.3 Fixed operating costs. In accordance with the methodology defined in section 2.4.2, is the estimation of the fixed operating costs partly a function of the estimated capital costs and partly directly dependent on the operating labour requirements. Reported data on the labour costs for the KRC plant in question indicate three personnel per shift. Based on the above estimate, the cost of direct labour is estimated to be:
  3 x 8760 x 26.25   = 689 850 ECU/y.

With the above data the cost estimation calculation can be carried out. The final results are summarised in Tables I, Ia and II. The Tables I and Ia give all cost items in detail as a demonstration of the applied calculation method. Table II and all the following tables give the data in a somewhat more condensed form.

4. Comparative cost study

The main purpose of the study by Pax was to provide a background for the appraisal of the economic outlooks of a new regenerative FGD process, the Ispra Mark 13A process.

The proposed procedure was to start with the development of a suitable analytical methodology for comparative cost evaluation and a preliminary evaluation of capital and operating costs for a number of current processes. In a second stage, the same economic analysis can be applied to the new process and, possibly, its most promising areas of application can be identified.

In the first phase, four processes were investigated:
  Wet limestone/gypsum (KRC)
  Wellmann-Lord
  Magnesium Oxide Process
  Bergbauforschung process.

The cost analysis was carried out for two capacities, 100 and 500 MWe, and for three different fuels, European hard coal, lignite and petroleum coke. During this phase, the cost module methodology was developed. The results indicated that the limestone/gypsum and the Wellmann-Lord process were the most important competitors.

In the second phase, a cost comparison between the KRC process, Wellmann Lord and the Ispra Mark 13A process was carried out. The main items of this study will be discussed below. The methodology applied to the limestone/gypsum process has been adequately described in Chapter 3.1.

Process	Base case	KRC	KRC	KRC	KRC
Fuel	Coal	Coal	Coal	Coal	Coal
Sulphur (%)	1.3	1.0	2.0	4.0	6.0
Capacity (Mwe)	423	500	500	500	500
FG Flow (10^6 m^3/h)	1.49	1.75	1.76	1.76	1.76
S removal rate (kg/h)	1547	1224	2658	5607	8648
CAPITAL COSTS (mio ECU)					
FG-module	24.46	27.23	27.33	27.33	27.33
S-module	21.80	18.65	31.28	51.46	68.70
TPC (= TFI)*	46.26	45.88	58.61	78.79	96.03
General facilities	*	*	*	*	*
Engineering	*	*	*	*	*
Project contingency	*	*	*	*	*
Process contingency	*	*	*	*	*
TFI	46.26	45.88	58.61	78.79	96.03
Royalty	0.46	0.46	0.58	0.79	0.96
Start-up costs	1.68	1.63	2.22	3.28	4.29
Working capital	0.41	0.37	0.63	1.18	1.74
IDC	4.62	4.59	5.86	7.88	9.60
TOTAL INVESTED CAPITAL	53.43	52.93	67.27	91.92	112.62

TABLE I
KRC process - 500 MWe Investment costs
*) = given capital costs are TFI costs

Process	Base case	KRC	KRC	KRC	KRC
Fuel	Coal	Coal	Coal	Coal	Coal
Sulphur (%)	1.3	1.0	2.0	4.0	6.0
Capacity (Mwe)	423	500	500	500	500
FG Flow (10^6 m^3/h)	1.49	1.75	1.76	1.76	1.76
S removal rate (kg/h)	1547	1224	2658	5607	8648
FIXED OPERATING COSTS (mio ECU/y)					
Direct labour	0.69	0.69	0.69	0.69	0.69
Maintenance	1.54	1.53	1.95	2.63	3.20
Indirect labour	0.39	0.39	0.44	0.52	0.59
Tax and insurance	1.51	1.47	1.87	2.52	3.07
TOTAL	4.13	4.08	4.95	6.36	7.55
VARIABLE OPERATING COSTS					
FG-module (mio ECU/y)	1.47	1.74	1.74	1.74	1.74
S-module (mio ECU/y)	3.45	2.73	5.88	12.41	19.14
TOTAL (mio ECU/y)	4.92	4.47	7.62	14.15	20.88
LEVELISED ANNUAL DCF CHARGE	5.14	5.10	6.48	8.85	10.85
TOTAL ANNUAL COSTS (mio ECU/y)	14.19	13.65	19.05	29.36	39.28
FG related power (MW)	3.3	3.9	3.9	3.9	3.9
S-related power (MW)	3.3	4.0	8.7	18.3	28.2
Total FGD power	6.6	7.9	12.6	22.2	32.1
Net station busbar power (MW)	416.4	492.1	487.4	477.8	467.9
FGD DCF cost (ECU mills/net kWh)	5.56	4.52	6.37	10.02	13.69

TABLE Ia
KRC process - 500 MWe
Operating costs

Capacity (Mwe)	Base case	100	100	100	100
Sulphur (%)	1.3	1.0	2.0	4.0	6.0
SO_2 in flue gas (ppm)	810	630	1250	2560	3780
S removal rate (kg/h)	1547	89	376	965	1573
FG Flow (10^6 m³/h)	1.49	0.35	0.35	0.35	0.35
CAPITAL COSTS					
FG-module (mio ECU)	24.46	9.63	9.63	9.63	9.63
S-module (mio ECU)	21.80	2.89	7.57	14.24	19.75
TPC (mio ECU)	46.26	12.52	17.19	23.87	29.38
Engineering, contingencies.....	*	*	*	*	*
TFI (mio ECU)	46.26	12.52	17.19	23.87	29.38
Royalties, start-up...	7.17	1.87	2.62	3.73	4.71
TOTAL INVESTED CAPITAL (mio ECU)	53.43	14.39	19.81	27.60	34.09
VARIABLE OPERATING COSTS					
FG-module (mio ECU/y)	1.47	0.34	0.34	0.34	0.34
S-module (mio ECU/y)	3.45	0.20	0.84	2.16	3.53
TOTAL	4.92	0.54	1.19	2.51	3.87
FIXED OPERATING COSTS (mio ECU/y)	4.13	1.73	2.04	2.49	2.86
LEVELISED ANNUAL DCF CHARGE	5.14	1.39	1.91	2.66	3.28
TOTAL ANNUAL COSTS (mio ECU/y)	14.19	3.66	5.14	7.66	10.01
FG related power (MW)	3.3	0.8	0.8	0.8	0.8
S-related power (MW)	3.3	0.3	1.2	3.2	5.2
Net station busbar power (MW)	416.4	98.9	98.0	96.0	94.0
FGD DCF cost (ECU mills/net kWh)	5.56	6.04	8.55	13.00	17.36

Table II
KRC process - 100 MW
*) = given capital costs are TFI costs

4.1 Input/output values

In the study a consistent set of values must be established for the unit costs of all plant inputs and outputs required for the calculation of the operating costs. The backgrounds for the values assigned to each item is discussed in detail in Ref. [10]. In the present paper, we give a summary of the values of the most important items:

Item	Value	Unit
Electric power	0.0725	ECU/kWh
LP steam	0.0107	ECU/kg
Natural gas	0.180	ECU/Nm^3
Heavy fuel oil	0.105	ECU/kg
Process water	0.120	ECU/ton
Cooling water	0.040	ECU/ton
Demi-water	0.600	ECU/ton
Limestone	20.0	ECU/ton
Soda ash	72.0	ECU/ton
Lime	50.0	ECU/ton
Hydrogen bromide	2670	ECU/ton
Standard solid waste	50.0	ECU/m^3
Labour	26.25	ECU/manhour
Sulphur credit	120.0	ECU/ton
Sulphuric acid credit	45.0	ECU/ton
Hydrogen credit	0.052	ECU/Nm^3

4.2 The Wellman-Lord process

The Wellman-Lord process is a wet regenerative FGD process based on the absorption of sulphur dioxide in a sodium sulphite solution. Regeneration of the formed sodium bisulphite yields a relatively concentrated sulphur dioxide gas stream, which can then be processed to sulphur or sulphuric acid. Process details are given in the relevant literature and in a separate paper of the present course and will not be discussed here.

A review of fullscale FGD plants suitable as a Base Case Process for the plant module method, indicates that there are few reference plants available in Europe. In Europe there were only two processes in operation in 1987. Moreover, there were no detailed cost data published in the open literature. In regard to data quality, the best documented material is found in the EPRI report series [8]. However, this information is not current (1982) and relates specifically to U.S. conditions. Notwithstanding, it was chosen to use the EPRI data as the Base Case of the Wellman-Lord process. The obtained data were checked with literature data for the process in operation at the Schwechat refinery in Austria [12] and found to be in acceptable agreement. The Base Case is a 2 x 500 MW power station fired with 4% sulphur coal. Total flue gas flow rate is 4.53 million m^3/h, SO_2 content 7.6 g/m^3, desulphurisation rate 90%.

4.2.1 Capital costs The original data for the base case process are summarised below:
- SO_2 removal system — 50 $/kW
- Regeneration system — 67 $/kW
- FG handling system — 30 $/kW

Reagent feed system	2 $/kW
Waste (sulphate) handling system	14 $/kW
Sulphur plant	17 $/kW

backgrounds for the values assigned to each item is discussed in detail in Ref. [10]. In the present paper, we give a summary of the values of the most important items:

Item	Value	Unit
Electric power	0.0725	ECU/kWh
LP steam	0.0107	ECU/kg
Natural gas	0.180	ECU/Nm^3
Heavy fuel oil	0.105	ECU/kg
Process water	0.120	ECU/ton
Cooling water	0.040	ECU/ton
Demi-water	0.600	ECU/ton
Limestone	20.0	ECU/ton
Soda ash	72.0	ECU/ton
Lime	50.0	ECU/ton
Hydrogen bromide	2670	ECU/ton
Standard solid waste	50.0	ECU/m^3
Labour	26.25	ECU/manhour
Sulphur credit	120.0	ECU/ton
Sulphuric acid credit	45.0	ECU/ton
Hydrogen credit	0.052	ECU/Nm^3

4.2 The Wellman-Lord process

The Wellman-Lord process is a wet regenerative FGD process based on the absorption of sulphur dioxide in a sodium sulphite solution. Regeneration of the formed sodium bisulphite yields a relatively concentrated sulphur dioxide gas stream, which can then be processed to sulphur or sulphuric acid. Process details are given in the relevant literature and in a separate paper of the present course and will not be discussed here.

A review of fullscale FGD plants suitable as a Base Case Process for the plant module method, indicates that there are few reference plants available in Europe. In Europe there were only two processes in operation in 1987. Moreover, there were no detailed cost data published in the open literature. In regard to data quality, the best documented material is found in the EPRI report series [8]. However, this information is not current (1982) and relates specifically to U.S. conditions. Notwithstanding, it was chosen to use the EPRI data as the Base Case of the Wellman-Lord process. The obtained data were checked with literature data for the process in operation at the Schwechat refinery in Austria [12] and found to be in acceptable agreement. The Base Case is a 2 x 500 MW power station fired with 4% sulphur coal. Total flue gas flow rate is 4.53 million m^3/h, SO_2 content 7.6 g/m^3, desulphurisation rate 90%.

4.2.1 Capital costs The original data for the base case process are summarised below:

SO_2 removal system	50 $/kW
Regeneration system	67 $/kW
FG handling system	30 $/kW
Reagent feed system	2 $/kW

Waste (sulphate) handling system 14 $/kW
Sulphur plant 17 $/kW
Particulate removal system 40 $/kW
General 3 $/kW

It has to be remarked that the particulate removal system (electrostatic precipitators) are in the present study not included in the FGD plant. The data are critically assessed, transformed from 1982 US $ to 1988 ECU and recalculated for one train 500 MW plant. The results are given below:

FG-module (2.26 million m³/h)
FG handling 6.20 mio ECU
SO₂ removal 14.27 mio ECU
General 1.31 mio ECU
Land 0.02 mio ECU
TOTAL 21.80 mio ECU

It follows that F_{FG} is equal to **12.66**.

S-module (7750 kg/h S)
Absorbent cycle 36.10 mio ECU
Sulphur plant 10.21 mio ECU
TOTAL 46.31 mio ECU

It follows that F_S = **0.118**

It has to be noted that the above figures are all on a TPC basis, so that additions for engineering, contingencies etc. have to be added according to the procedures described in Chapter 2. The Wellman-Lord process is considered as an industrially proven process. This has its consequence for the choice of the cost factor for the process contingency. The adopted value is in this case: 0.1 x TPC. Project contingency is maintained at 0.2 x TPC.

4.2.2 Variable operating costs Based on the data given in the Base Case study, the following reagent and utilities requirements on a unit basis were calculated:

FG-module (per million m³ of flue gas)

Electric power, FG fans	2.25	MW	163.1	ECU
SO₂ removal	0.556	MW	40.3	ECU
Steam (reheat)	19.9	kg	0.2	ECU
Standard solid waste	1.0	m³	50.0	ECU
TOTAL			253.6	ECU/million m³

S-module (per kg of S removed)

Soda ash	0.09	kg	0.006	ECU
Natural gas fuel	0.16	Nm³	0.029	ECU
Natural gas reductant	0.46	Nm³	0.083	ECU
Electric power	1.39	kW	0.101	ECU
Net steam used	16.14	kg	0.140	ECU
Boiler feed water	1.93	l	0.001	ECU
Process water	17.8	l	0.002	ECU
Cooling water	300	l	0.012	ECU

Sulphur credit	1.0	kg	0.120 ECU
TOTAL			**0.254 ECU/kg S**

4.2.3 Fixed operating costs In the applied cost factor methodology, the fixed operating costs are partially dependent on the capital costs and partially on the operating labour requirements. It is estimated that for a full Wellman-Lord process the following personnel is required:

Plant shift supervisor:	1
Main FGD area operative:	1
Regeneration area operative:	1
Sulphur plant operative:	1
TOTAL per shift:	4

Then the annual cost of direct labour is estimated to be:
  4 x 8760 x 26.25   =   919 800 ECU/y

Based on the above figures and using the cost factor and the process plant module technology, the cost estimations for the Wellman-Lord process can be calculated. The results are summarised in Tables III and IV.

4.2 The Ispra Mark 13A process

The Ispra Mark 13A process was invented at the Joint Research Centre of the European Community located in Ispra, Italy. It is a regenerative wet scrubbing process, employing as scrubbing liquid an acid aqueous solution containing sulphuric acid, hydrobromic acid and a small percentage of bromine. The sulphur dioxide is converted into sulphuric acid and the bromine used in this reaction is regenerated by electrolysis. Technical details of the process are given in a separate paper. The process has reached the pilot plant stage. An installation with a capacity of 32 000 Nm³/h of flue gas from a power station fired by a mixture of fuel oil and refinery gas has been built near Cagliari in Sardinia, Italy. The maximum sulphur dioxide content of the flue gas is 4.5 g/m³.
The Base Case Process consists of a detailed preliminary offer, jointly prepared by the engineering firm Kraftanlagen Heidelberg and the JRC Ispra [13]. The plant in question was a FGD with a capacity of 287 000 Nm³/h of flue gas and a SO_2 content of 0.2 vol%.
The plant cost module concept for the Ispra Mark 13A process had to be slightly modified, particularly for the S-module. The capital costs for each module is normally considered to follow the two third power rule. However, in the Ispra process the regeneration takes place by electrolysis. The cost of electrolysis electrodes is a major item in the module cost and this cannot be scaled by the 2/3 power rule. The cost of electrodes is practically proportional to the plant capacity. Therefore, a split in the S-module capital costs is made: 60% of the electrolyser costs are considered to be proportional to the plant capacity, and the remaining 40% is scaled by the $\frac{2}{3}$ power rule.

4.3.1 Capital costs The data from Ref.[13] referring to the FG-module are the following:

Flue Gas Fan	0.80 mio DM
Regenerative Heat Exchanger	2.80 mio DM
Reactor plus Scrubber	1.39 mio DM

Capacity (Mwe)	Base case	500	500	500	500
Sulphur (%)	4.0	1.0	2.0	4.0	6.0
SO_2 in flue gas (ppm)	2667	630	1250	2560	3780
S removal rate (kg/h)	7750	1224	2658	5607	8648
FG Flow (10^6 m³/h)	2.26	1.75	1.76	1.76	1.76
CAPITAL COSTS					
FG-module (mio ECU)	21.80	18.39	18.46	18.46	18.46
S-module (mio ECU)	46.31	13.51	22.65	37.25	49.73
TPC (mio ECU)	68.11	31.90	41.11	55.71	68.19
Engineering, contingencies.....	35.76	16.75	21.58	29.25	35.80
TFI (mio ECU)	103.87	48.65	62.69	84.96	103.99
Royalties, start-up...	16.80	7.49	9.76	13.55	16.92
TOTAL INVESTED CAPITAL (mio ECU)	120.67	56.14	72.45	98.51	120.91
VARIABLE OPERATING COSTS					
FG-module (mio ECU/y)	3.51	2.72	2.74	2.74	2.74
S-module (mio ECU/y)	12.07	1.91	4.14	8.73	13.47
TOTAL	15.58	4.63	6.88	11.47	16.21
FIXED OPERATING COSTS (mio ECU/y)	8.34	4.55	5.51	7.04	8.35
LEVELISED ANNUAL DCF CHARGE	11.64	5.42	6.99	9.51	11.67
TOTAL ANNUAL COSTS (mio ECU/y)	35.56	14.60	19.38	28.02	36.23
FG related power (MW)	6.4	4.9	4.9	4.9	4.9
S-related power (MW)	10.8	5.7	12.4	26.1	40.2
Net station busbar power (MW)	482.8	489.4	482.7	469.0	454.9
FGD DCF cost (ECU mills/net kWh)	12.01	4.87	6.55	9.74	12.99

Table III
Wellman-Lord process, 500 MWe

Capacity (Mwe)	Base case	100	100	100	100
Sulphur (%)	4.0	1.0	2.0	4.0	6.0
SO$_2$ in flue gas (ppm)	2667	630	1250	2560	3780
S removal rate (kg/h)	7750	89	376	965	1573
FG Flow (10^6 m^3/h)	2.26	0.35	0.35	0.35	0.35
CAPITAL COSTS					
FG-module (mio ECU)	21.80	6.22	6.22	6.22	6.22
S-module (mio ECU)	46.31	2.32	6.09	11.47	15.91
TPC (mio ECU)	68.11	8.55	12.32	17.69	22.13
Engineering, contingencies.....	35.76	4.50	6.47	9.29	11.62
TFI (mio ECU)	103.87	13.04	18.79	26.98	33.75
Royalties, start-up...	16.80	1.95	2.77	4.02	5.07
TOTAL INVESTED CAPITAL (mio ECU)	120.67	14.99	21.57	31.00	38.82
VARIABLE OPERATING COSTS					
FG-module (mio ECU/y)	3.51	0.55	0.55	0.55	0.55
S-module (mio ECU/y)	12.07	0.13	0.58	1.50	2.46
TOTAL	15.58	0.68	1.13	2.05	3.01
FIXED OPERATING COSTS (mio ECU/y)	9.34	2.07	2.45	3.00	3.45
LEVELISED ANNUAL DCF CHARGE	11.64	1.44	2.08	2.99	3.74
TOTAL ANNUAL COSTS (mio ECU/y)	35.56	4.19	5.66	8.04	10.19
FG related power (MW)	6.4	1.0	1.0	1.0	1.0
S-related power (MW)	10.8	0.4	1.7	4.5	7.3
Net station busbar power (MW)	482.8	98.6	97.3	94.5	91.7
FGD DCF cost (ECU mills/net kWh)	12.01	6.93	9.49	13.86	18.13

Table IV
Wellman-Lord 100 MWe

Spray Nozzles	0.13 mio DM
Demisters	0.25 mio DM
Tanks and pumps	0.58 mio DM
Pipework, ducting and utilities	4.70 mio DM
Civil work	0.57 mio DM
Instrumentation and electrics	1.57 mio DM
Installation	0.57 mio DM
TOTAL	13.36 mio DM

Correcting for inflation and conversion to ECU yields 6.42 mio ECU for 0.287 mio Nm³/h of flue gas. The corresponding value for **F_{FG} is then 14.76**

The S-module is based on a removal rate of 723 kg/h of sulphur. A price breakdown is given below:

Electrolysis	6.80 mio DM
Columns, packings, nozzles ...	0.87 mio DM
Pumps and reheating	1.40 mio DM
Pipework, ducting and utilities	3.00 mio DM
Civil work	0.82 mio DM
Instrumentation and electrics	2.27 mio DM
Installation	0.82 mio DM
TOTAL	15.98 mio DM

Correcting for inflation, conversion to ECU and splitting into a linear and a 2/3 power scale up part gives the following results:

Electrodes (linear scale up)	1.95	mio ECU	-	$F_{S1} = 0.0027$ ECU/kg
Balance of S-module	5.74	mio ECU	-	$F_{S2} = 0.0712$

Given the fact that the Ispra Mark 13A process is only in the pilot plant stage, higher values for the project and process contingencies have to be applied than for industrially mature processes. According the guidelines of EPRI, for industrial processes a project contingency factor of 0.20 has been applied, whereas this factor for Ispra Mark 13A is brought to 0.30. Likewise, also the factor for process contingency has been increased from 0.10 to 0.25.

4.3.2 Variable operating costs Based on the data given in the Base Case study, the following reagent and utilities requirements were calculated on a unit basis:

FG-module (per million m³ of fluegas)

Electric power, FG fans	2.46	MW		178.4 ECU
Pumping duties	0.50	MW		36.1 ECU
Demi water	55.0	m³		33.0 ECU
HBr losses	5.0	kg		13.4 ECU
TOTAL				**260.9 ECU/million m³**

S-module (per kg of S removed)

Electric power	2.84	kW		0.206 ECU
Electrode wear	---			0.052 ECU
Sulphuric acid credit (96%)	3.19	kg	-	0.144 ECU
Hydrogen credit	0.25	m³	-	0.013 ECU
TOTAL				**0.101 ECU/kg S**

4.3.3 Fixed operating costs The fixed operating costs are partially dependent on the capital costs and partially on the operating labour requirements. It is estimated that a full scale commercial unit can be operated with 3 personnel per shift.
Then the annual cost of direct labour is estimated to be:
3 x 8760 x 26.25   =   0.690 mio ECU/y

Based on the above figures and using the cost factor and the process plant module technology, the cost estimations for the Ispra Mark 13A process can be calculated. The results are summarised in Tables V and VI.

5. Results

The best single-valued measure of the cost effectiveness of FGD processes is the total annual DCF costs per net kWh. These costs are summarised below for the case of a 500 MW power station (all values in ECU mills/kWh or ECU/MWh):

	KRC process	Wellman-Lord	Ispra Mark 13A
1.0% S	4.52	4.87	4.88
2.0% S	6.37	6.55	6.18
4.0% S	10.02	9.74	8.67
6.0% S	13.69	12.99	11.18

For a 100 MW power station the values are considerably higher:

1.0% S	6.04	6.93	7.02
2.0% S	8.55	9.49	8.92
4.0% S	13.00	13.86	12.19
6.0% S	17.36	18.13	15.33

There is a discontinuity in the case for 100 MW due to the step change of the allowable sulphur dioxide emission concentrations in accordance with the EC Directive. The admissible value decreases from 2.0 to 1.2 g/m^3 at 100 MWe and from 1.2 g to 0.25 g/m^3 at 300 MW. This has its influence on the required degrees of desulphurisation and, consequently, on the various process costs.

The aforementioned data are also represented graphically in Figs. 1 and 2. In Fig. 1 an average curve for FGD costs given by Ando [2] is included. The agreement of both data sets seems to be acceptable, although in the present study the effect of the sulphur dioxide concentration on the costs is higher than given in Ref. [2].

It can be concluded that for low capacities and for low sulphur contents, the market chances for new, regenerative processes are low. The traditional lime/limestone processes seem to be more attractive in this region. When the sulphur concentrations become more important, there is a distinct advantage for the regenerative processes.

Capacity (Mwe)	Base case	500	500	500	500
Sulphur (%)	--	1.0	2.0	4.0	6.0
SO_2 in flue gas (ppm)	2000	630	1250	2560	3780
S removal rate (kg/h)	723	1224	2658	5607	8648
FG Flow (10^6 m³/h)	0.287	1.75	1.76	1.76	1.76
CAPITAL COSTS					
FG-module (mio ECU)	6.42	21.43	21.52	21.52	21.52
S-module (mio ECU)	7.69	11.45	20.84	37.62	53.41
TPC (mio ECU)	14.11	32.88	42.36	59.14	74.93
Engineering, contingencies.....	10.94	25.48	32.83	45.83	58.07
TFI (mio ECU)	26.05	58.37	75.19	104.97	133.00
Royalties, start-up...	3.78	7.99	10.99	15.32	19.43
TOTAL INVESTED CAPITAL (mio ECU)	29.83	66.35	86.18	120.29	152.43
VARIABLE OPERATING COSTS					
FG-module (mio ECU/y)	0.46	2.81	2.82	2.82	2.82
S-module (mio ECU/y)	0.45	0.76	1.65	3.47	5.36
TOTAL	0.91	3.57	4.47	6.29	8.18
FIXED OPERATING COSTS (mio ECU/y)	2.52	2.04	5.68	7.57	9.35
LEVELISED ANNUAL DCF CHARGE	2.88	6.54	8.32	11.61	14.71
TOTAL ANNUAL COSTS (mio ECU/y)	6.31	14.71	18.47	25.47	32.24
FG related power (MW)	3.0	5.2	5.2	5.2	5.2
S-related power (MW)	2.0	3.5	7.5	15.9	24.6
Net station busbar power (MW)	76.0	491.3	487.3	478.9	470.2
FGD DCF cost (ECU mills/net kWh)	13.54	4.88	6.18	8.67	11.18

Table V
Ispra Mark 13A - 500 MWe

Capacity (Mwe)	Base Case	100	100	100	100
Sulphur (%)	--	1.0	2.0	4.0	6.0
SO_2 in flue gas (ppm)	2000	630	1250	2560	3780
S removal rate (kg/h)	723	89	376	965	1573
FG Flow (10^6 m³/h)	0.287	0.35	0.35	0.35	0.35
CAPITAL COSTS					
FG-module (mio ECU)	6.42	7.33	7.33	7.33	7.33
S-module (mio ECU)	7.69	1.66	4.72	9.56	13.90
TPC (mio ECU)	14.11	8.99	12.05	16.90	21.23
Engineering, contingencies.....	10.94	6.97	9.34	13.10	16.45
TFI (mio ECU)	26.05	15.95	21.38	29.99	37.68
Royalties, start-up...	3.78	2.27	3.01	4.20	5.27
TOTAL INVESTED CAPITAL (mio ECU)	29.83	18.22	24.39	34.19	42.95
VARIABLE OPERATING COSTS					
FG-module (mio ECU/y)	0.46	0.56	0.56	0.56	0.56
S-module (mio ECU/y)	0.45	0.06	0.23	0.60	0.97
TOTAL	0.91	0.62	0.79	1.16	1.53
FIXED OPERATING COSTS (mio ECU/y)	2.52	1.88	2.21	2.74	3.22
LEVELISED ANNUAL DCF CHARGE	2.88	1.75	2.35	3.29	4.14
TOTAL ANNUAL COSTS (mio ECU/y)	6.31	4.25	5.35	7.19	8.88
FG related power (MW)	3.0	1.0	1.0	1.0	1.0
S-related power (MW)	2.0	0.3	1.1	2.8	4.5
Net station busbar power (MW)	76.0	98.7	97.9	96.2	94.5
FGD DCF cost (ECU mills/net kWh)	13.54	7.02	8.92	12.19	15.33

Table VI
Ispra Mark 13A - 100 MWe

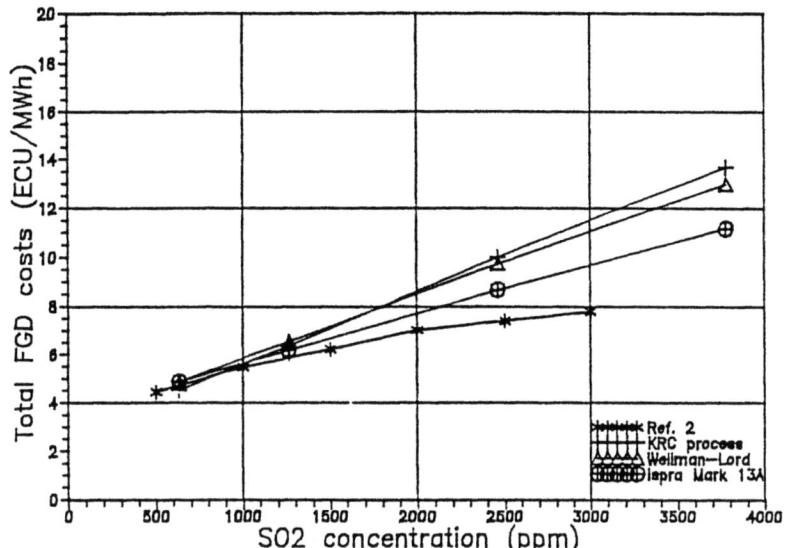

Fig. 1
Total annual (DCF) FGD costs
500 MWe power station

This is mainly due to the fact that the variable operating costs for this type of processes are considerably lower (by-product credit!). This feature is illustrated in Fig. 3, where the variable operating costs are given as a function of sulphur dioxide concentration in the flue gas. It follows from the figure that the variable operating costs for Ispra Mark 13A are considerable lower (up to 50 - 60%) than those for the other two processes.

These lower operating costs are counterbalanced by higher investment costs. This follows clearly from Fig. 4, where the specific investment costs (TFI), expressed in ECU/kW, are given as a function of sulphur dioxide concentration in the flue gas. Here, the opposite of the situation with the variable operating costs occurs, the Ispra Mark 13A process has considerably higher costs than the other processes. These higher investment costs are for a large part caused by the presence of an electrolyser section, which may represent up to 25% of the total investment.

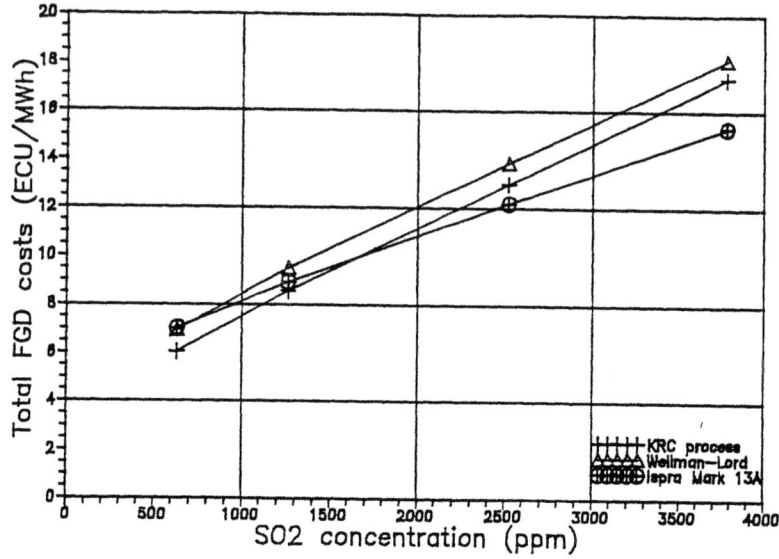

Fig. 2
Total annual FGD costs (DCF)
100 MWe power station

6. Denoxing

The present paper is concentrated on the comparative cost estimation for FGD processes and the development of the plant cost module technique.
A similar study for denoxing and combined desulphurisation/denoxing processes is as yet outstanding. It is felt that the application of the plant cost module technique in these cases will be particularly helpful, as it allows the splitting of the plant investment costs as well as the operating costs into a FG module, a S-module and a N-module.

In the SCR process the investment cost elements which have to be attributed to the N-module are:
 - costs of the catalyst
 - equipment costs
 - ammonia storage and supply

To the FG-module belong the costs for:
 - additional piping
 - reheating of the flue gas.

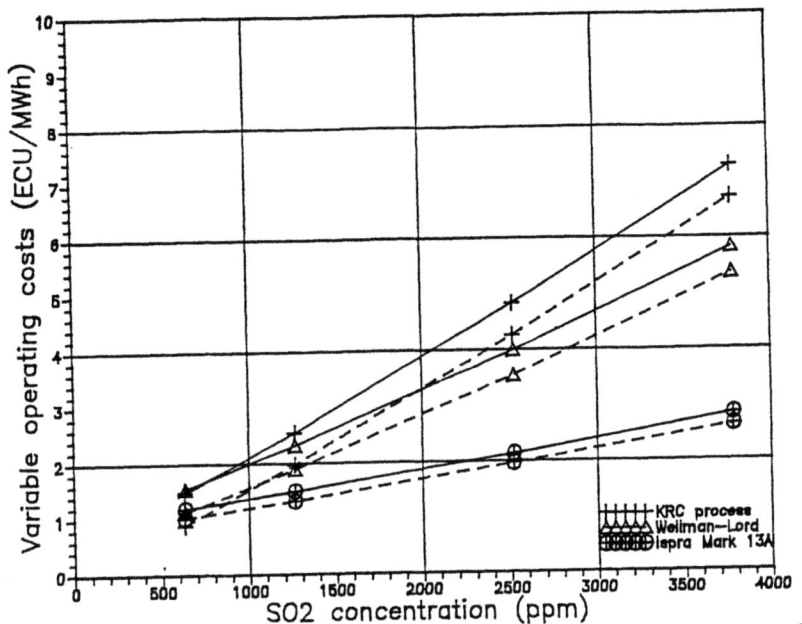

Fig. 3
Variable operating costs
Dashed line = 100 MW, full line = 500 MW

Variable operating costs can be splitted up in the same way.
In this way, a clear distinction between the High Dust and the Low Dust variation of the process can be made. Also a fair comparison with combined proceses, like the Bergbauforschung/Uhde process or the EBDS process can be carried out using this methodology.
Such a study is in preparation and results are expected to be available in a few months.
For the moment, the reader has to rely on the literature data given by Ando and Schärer & Haug [2, 3], i.e. investment costs between 40 and 57 ECU/kWe and total denoxing costs (DCF) between 2.3 and 3.6 ECU/MWh.

7. Conclusions

In the present paper a methodology for a comparative cost analysis is presented, based on the splitting of the process costs into various cost modules.
The proposed methodology seems to be attractive and will also be applied in a comparative cost study for denoxing processes.

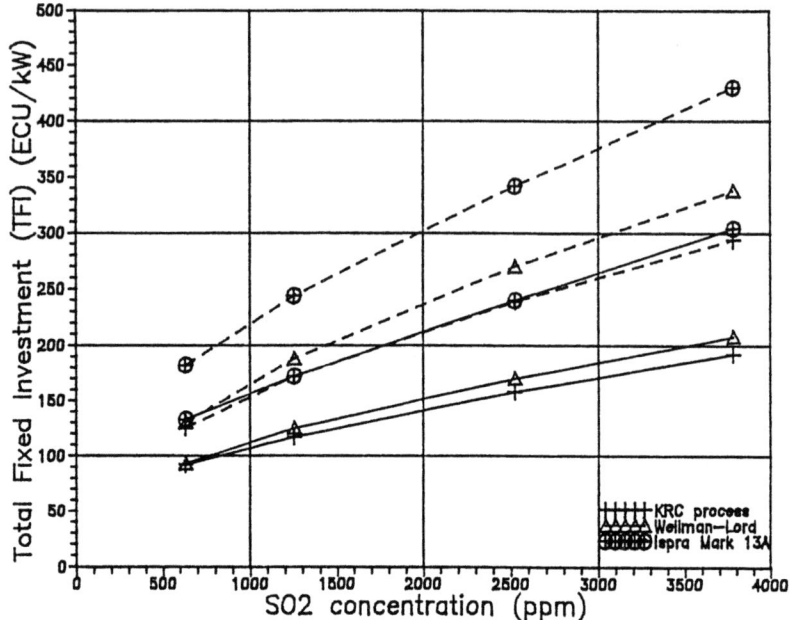

Fig. 4
Total fixed Investment costs (TFI)
Dashed line = 100 MW, full line = 500 MW

An example of the application of this technique is given where it was pointed out that the possible future of new regenerative FGD processes is in the region of high sulphur concentrations in the flue gas and rather large plant capacities.

A final remark must be made:
The results of any study based on the cost module concept, depend completely of the quality of the Base Case Data. These data are sometimes referring to fully developed industrial processes, sometimes they are extrapolations from U.S. conditions and they may even be based on pilot and demonstration plant results. It must therefore be borne in mind that comparative cost studies can only represent an indication of the trends and will never be more accurate than 20%. Conclusions must always been drawn with great care.

8. References

[1]   J. Jung, "Investitionsaufwand für die SO_2 - und NO_x-Minderung in der deutschen Elektrizitätswirtschaft". VGB Kraftwerkstechnik (1988), no. 2 pp. 154-157

[2]   J. Ando, "Recent Developments in SO_2 and NO_x Abatement Technology for Stationary Sources", in L.J. Brasser and W.C. Mulder (eds.) "Man and his Ecosystem, Proceedings of the 8th World Clean Air Congress 1989" Elsevier Amsterdam 1989, Vol. 4, pp 129-140

[3]   B. Schärer and N. Haug,
Staub - Reinhaltung der Luft 50 (1990) 139-144

[4]   F.P. Boer, L.L. Hegedus, T.R. Gouker and K.P. Zak,
Chemtech May 1990 312-319

[5]   E.A. Stallworthy, Chem. Eng., June 1970

[6]   A.J. Bogers, "Waterstof als Energiedrager"TNO Report, TNO, The Hague , September 1975, pp. 67-68

[7]   R.W.Scheck et. al. "Economics of Four FGD Systems", EPRI-CS-1677, Research Project 1180-3, prepared by Stearns-Roger Engineering Corporation, Denver Colorado for EPRI, Palo Alto, January 1981

[8]   R.J. Keeth and P.A. Ireland, "Economic Evaluation of FGD Systems" Volumes 1-4, CS-3342, Research Project 1610-2, prepared by Stearns Catalytic Corporation, Denver Colorado for EPRI, Palo Alto, July 1985

[9]   P.W. Dacey and D.R. Cope, ""Flue gas desulphurisation - system performance", IEA Coal Research, London , 1986

[10]   Pax Technology Transfer Ltd, "Flue Gas desulphurisation - Comparative Cost Study for Four Regenerative Processes", Contract Expect 9 for DG XIII/C of European Commission, Luxemburg, September 1989

[11]   P. Necker and J.H. Strauss, "Function and Costs of a flue gas desulphurisation plant on the basis of a limestone/gypsum wet process, using the example of Unit 5 of Neckarwerke, Altbach, Deizisau Power Station", Proceedings of Workshop on Emission Control Costs, Esslingen am Neckar 1987. Institute for Industrial Production for Executive Body for the Convention of Long-Range Transboundary Air Pollution, Karlsruhe.

[12]   CONCAWE Air Quality Management Group's Special Task Force of Flue Gas Desulphurisation Costs; Report No. 3/88 "Regenerative Flue Gas Desulphurisation in European Oil Refineries - Cost Estimates based on a European Application", CONCAWE, The Hague, 1988

[13]   "Preliminary Budget Price Proposal for AGIP Petroli based on installation in the Po refinery at Sannazzaro de Borgondi (Pavia) Italy" prepared by Kraftanlagen Heidelberg and JRC Ispra, Heidelberg, June 1988.

SITUATION IN THE UNITED STATES AND JAPAN

G. CAPRIOGLIO
FERLINI/GENERAL ATOMICS DEVELOPMENT CORPORATION
PO Box 85608
San Diego, California
USA

1. Introduction

More than half of the Flue Gas Desulfurization capacity in operation around the world is installed in the United States. The present market for Flue Gas Desulfurization plants in the United States is less than 10% of the worldwide market. The market for Flue Gas Desulfurization plants in the United States in 1996 will be at least 50% of the worldwide market. This is, in a nutshell, the status of the FGD industry in the United States: very active during the 70's, slow in the 80's and preparing for a boom in the 90's.

The controlling factor for this discontinuity has been the long gestation of the federal legislation on air pollutants emission from power plants. In Japan, where Flue Gas Desulfurization systems have begun to appear at the same time than in the United States (early 70's), the demand for new systems has continued in unison with the installation of new coal fired power plants.

Control of nitrogen oxides in industrial waste gases has received a substantially different attention in the United States and in Japan. In the United States the relatively high allowable limits have resulted in combustion modification as the primary system for NO_x control, while in Japan flue gas treatment is the preferred method.

From the point of view of technology development, for both sulfur dioxide and nitrogen oxides control, a large factor has been the American Clean Coal Technology Program. Organized and managed by the U.S. Department of Energy, the Clean Coal Technology Program has provided and is providing to domestic and foreign industry a substantial financial support for the demonstration of the commercial viability of a variety of new technologies. The impact of this program on the evolution of the technology for the control of flue gas emissions is going to be felt by the industry during the next decade.

We will examine some of the technical and legislative issues which are influencing the situation of flue gas emission control in the United States and Japan, and we will mainly review the status of the U.S. utility industry during the next decade, since in absolute terms this arena is where the largest game is going to be played.

2. Coal and the Utility Industry

In both the United States and in Japan, the power generation industry operates on defined geographical areas. In Japan, there are nine investor-owned companies supplying almost all of the nation's power in a consistent manner from power generation to transmission and to distribution. In the United States, the industry is much more subdivided with hundreds of investor-owned companies, public utilities and municipalities.

Coal represents today only 10% of the total electricity generated in Japan, but it is expected that its share will increase to 15% during the next decade. Figure 1 shows the generating capacity and the electric power generation for Japan in the year 1988 and 1998 [1].

The emission standards for coal-fired power plants having a flue gas rate higher than 200,000 m^3/h (corresponding to approximately 50 MW(e) of generating capacity) are respectively 50 and 100 mg/m^3 of NO_x for new and existing boilers and are locally set for SO_2. Figure 2 shows the average amount of SO_2 in the desulfurized flue gas in Japan as a function of the years, expressed as percentage of sulfur that the coal would have to contain in order to achieve such a level without desulfurization. The very low level of 0.1% corresponds to approximately 70 ppm of SO_2 in the flue gas.

No substantial changes are expected in the Japanese regulations, but since the overall environmental NO_x concentration has not improved during the past few years, probably because of the increase in the number of diesel powered vehicles, a more stringent NO_x control at fossil powered power plants might be required. In general, we can predict little change to the trend of the Japanese SO_2 and NO_x control industry during the next decade.

In the United States coal represent almost half of the generating capacity (approximately 300 GW out of the total 675 GW) [2]. The projections for the next decade indicate that coal will have more than half share of the approximately 50 GW of additional capacity (but this figure can easily become as high as 200 GW, mostly in coal plants).

Figure 3 shows the generating capacity by energy source for the United States in 1988 and Figure 4 the geographical distribution of the generating capacity. The coal-fired generating capacity is shown in Figure 5. Since coal is an abundant natural resource in the United States, it is also interesting to see how it is distributed (Figure 6) and how the nation is essentially divided into two parts from the point of view of the coals' sulfur content (Figure 7) [3].

A simple observation of these figures shows how the SO_2 control issue is predominantly strong in the eastern states, where large quantities of high-sulfur coal are produced and burned.

3. The Regulatory Process

Regulations on sulfur dioxide, nitrogen oxides and dust emission are usually fairly complex and site dependent, but some basic figures can give an idea of how stringent the laws of any given country are. In Table 1 we show a comparison of U.S. and German regulations for coal power plants.

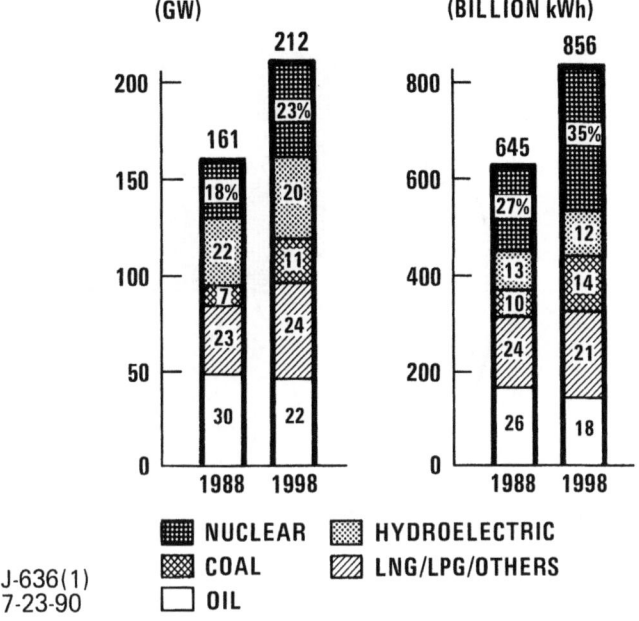

Figure 1. Generating capacity and electric power generation in Japan

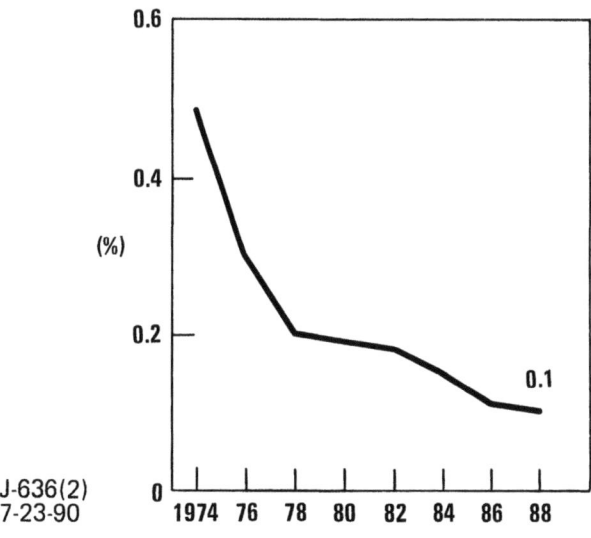

Figure 2. Sulfur content of fuels (in desulfurized flue gas)

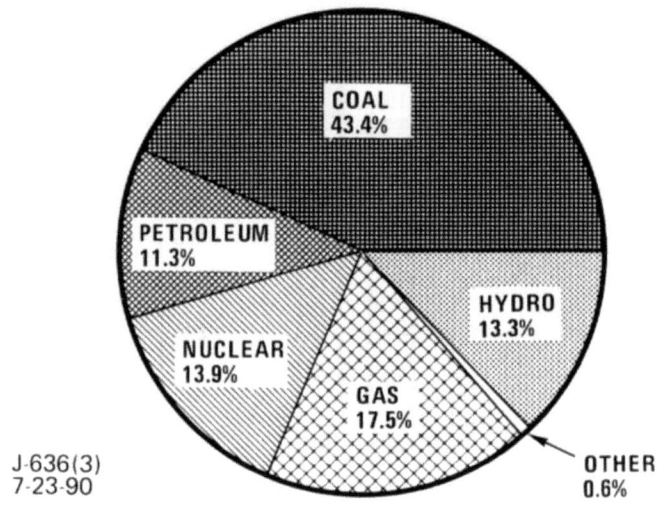

Figure 3. U.S. generating capability by energy source (1988)

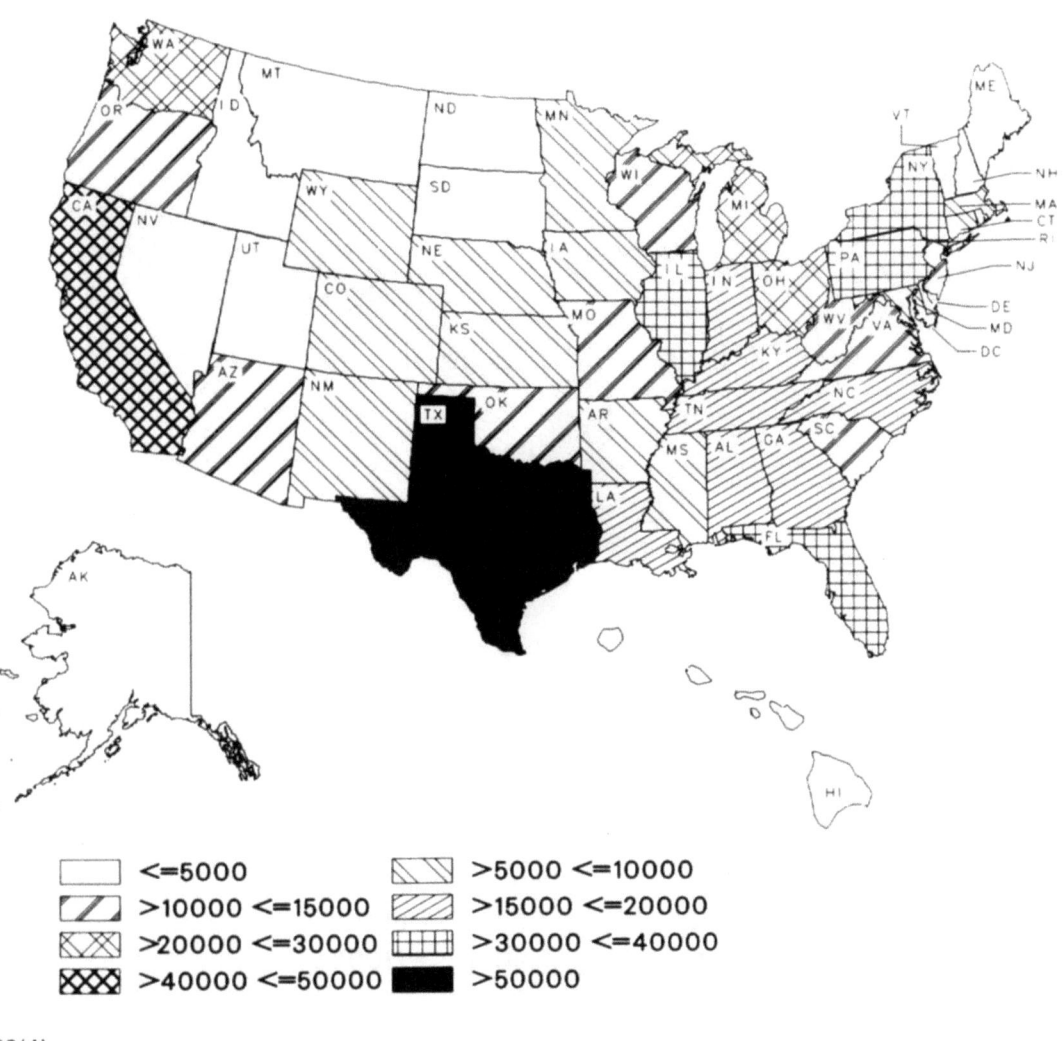

Figure 4. U.S. generating capability (1988) Megawatts

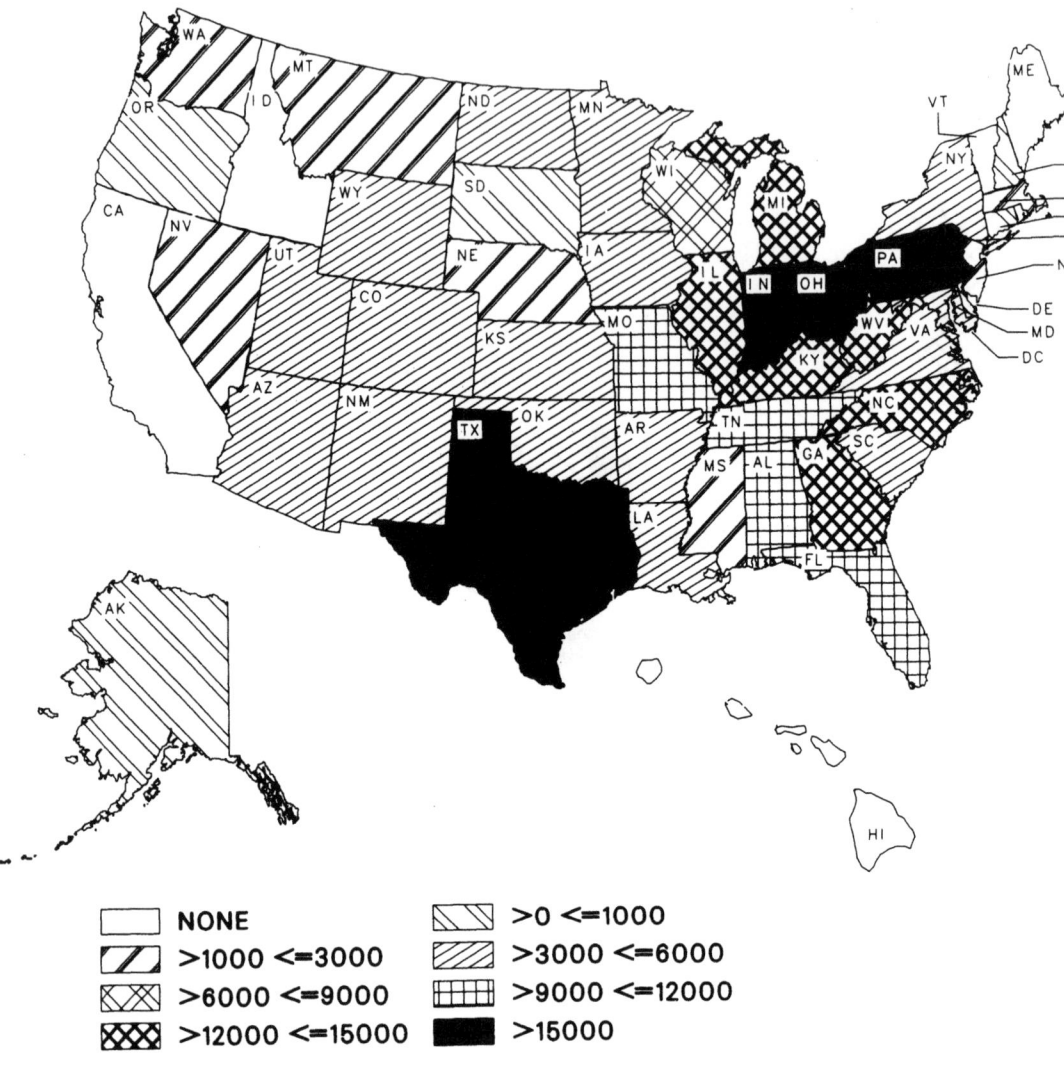

Figure 5. Coal-fired generating capacity (1988) (Megawatts)

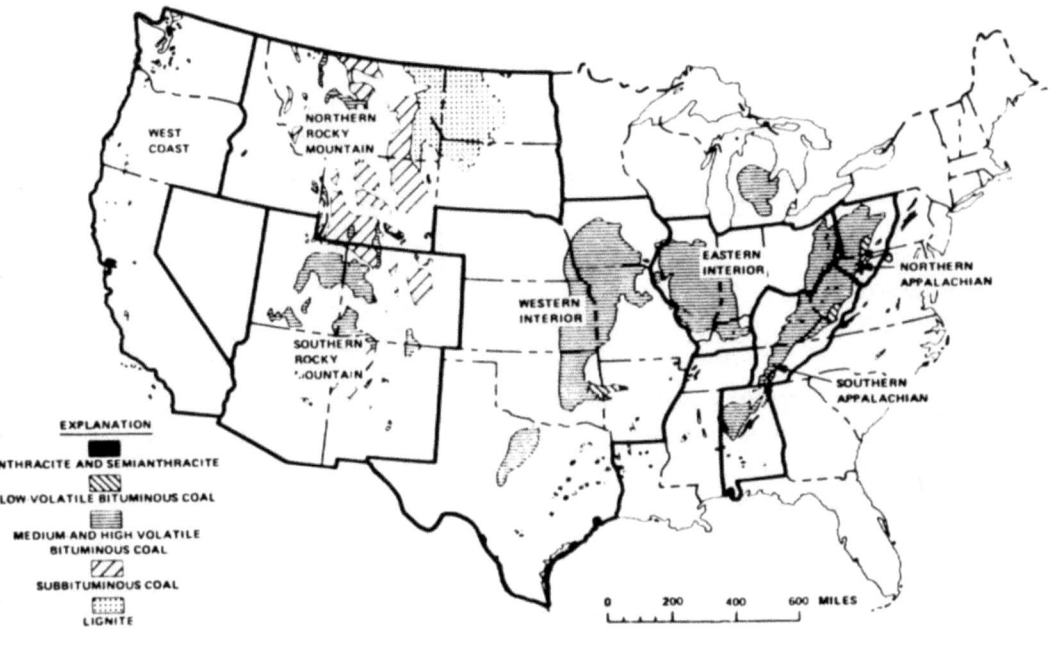

Figure 6. Coal producing regions

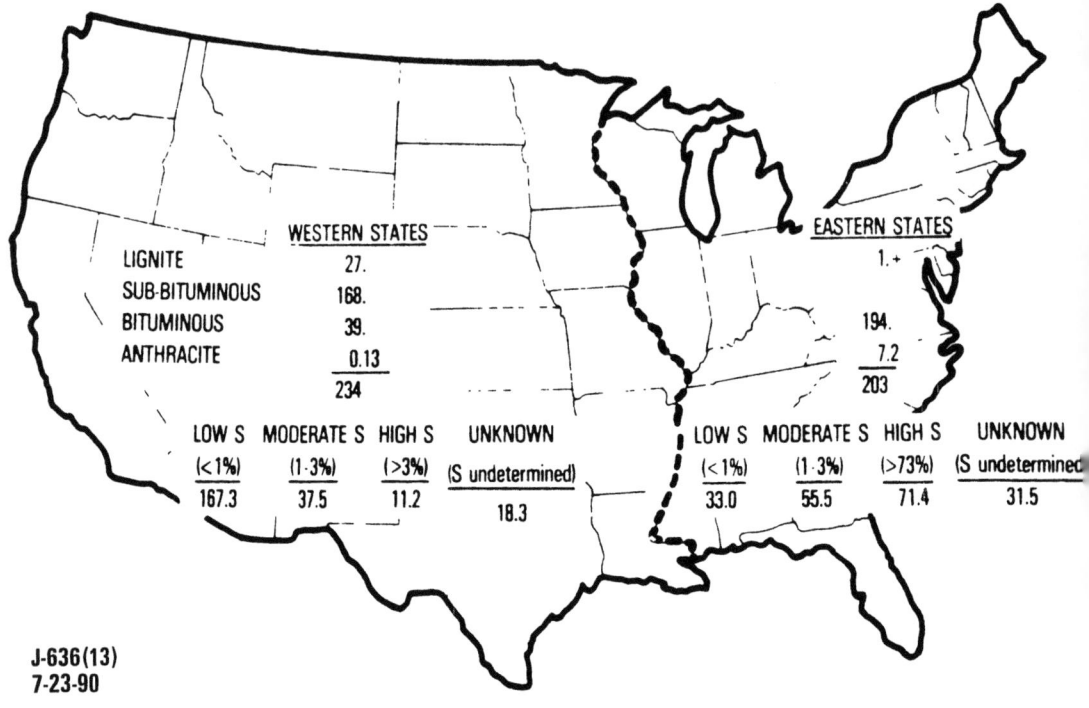

Figure 7. The reserve base by coal type and sulfur categories (in billions of tons)

TABLE 1.
COAL POWER PLANT EMISSION LIMIT COMPARISON

	Plant Size	Emission Limits (mg/m^3)	
		USA	FRG
SO$_2$	Small	1250	2000
	Large	1250/620	400
NO$_x$	Small	715	500
	Large	620	200
Dust	Small	100	150
	Large	30	50

Several remarks on the basic numbers shown in the table are necessary:
1) The Separation between small and large plants is consistently defined as 73 MW in the United States and is 5 MW for dust, and 300 MW for SO$_2$ and NO$_x$ in Germany, where also exist some intermediate provisions.
2) The standards for large plants in the United States apply only to boilers constructed after 1978 and after 1971 for small plants.
3) In Germany, SO$_2$ emission is further controlled by the sulfur emission ratio expressed as a percentage of total fuel sulfur content, while in the United States, for large plants, a reduction of 90% is generally required or a 70% reduction only when emissions are less than 620 mg/m^3 on a 30 days average.

All this is on going to change in the United States. The House and Senate are finalizing a legislation known as the Clean Air Act, which includes different standards for SO$_2$, NO$_x$ and dust and most importantly, requires all plants, including the old ones, to comply. It is expected that the new Clean Air Act will require all plants to comply in two phases, defined by various criteria, Phase I by 1995 and Phase II by 2000.

The expected SO$_2$ emission standards for these retrofit plants are shown in Figure 8 in comparison with the standard for high sulfur and low sulfur U.S. plants and with the standards of other countries. The data are expressed in mg/Nm^3 to facilitate the comparison, but it must be kept in mind that all the U.S. regulations express the limits in pounds of pollutant per million Btu of thermal energy produced.

One of the objectives of the Clean Air Act is to reduce the annual SO$_2$ emission in the United States from 20 million tons to 10 million tons, and the new standards are aimed at achieving this results. A very controversial aspect of the new legislation is the provision that allows utilities to trade emission rights. A utility would be allowed to desulfurize in excess of

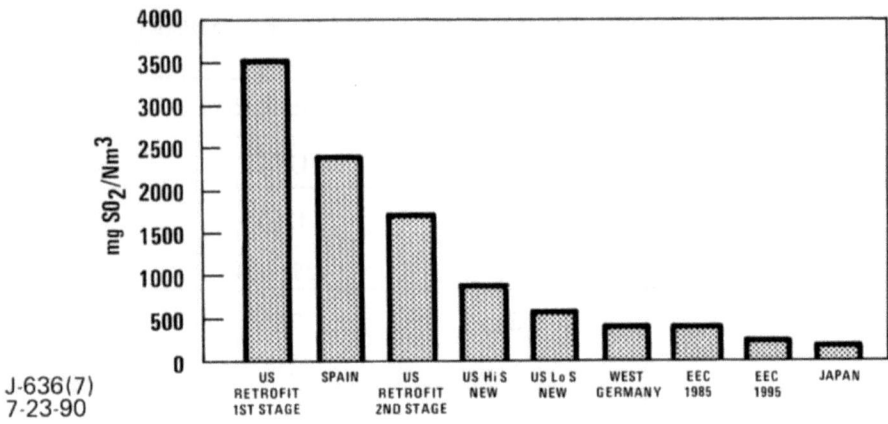

Figure 8.   Regulations based on outlet emissions in mg/Nm^3

the required limit and "trade" the extra amount of SO_2 removed with another utility that chooses not to desulfurize. Economists are already busy estimating the value of a ton of SO_2 removed as a tradeable commodity. Proponents of this provision say that this is as good as a tax break for the 107 plants with excess SO_2 emission. They can reduce emission below their required limit and sell the difference for a profit since the incremental cost is going to be less than what another utility would pay for avoiding building a desulfurization plant or switching to low-sulfur coal. Opponents to the emission trading are concerned about the different treatment that this will give to various utilities. Each utility will have a total emission cap: the emission cap is another provision in the Act requiring that new plants will not add additional SO_2 emission. A utility with a large coal fired base would be able to expand by desulfurizing below the limits on the retrofits and then using the emission rights to maintain the cap. By contrast, independent power companies who have no such emissions to bank could be restricted from building.

How the utility are going to comply with the requirements of the new legislation is very much a subject of speculations. Scrubbers, fuel- switching and plant decommissioning are feasible alternatives. A recent marketing study conducted by McIlvaine, a consulting company in Illinois [4], has surprisingly shown that only a relatively small number of plants will permanently switch to low-sulfur coal. Their conclusion is that many utilities initially considering full-switching are now planning to install scrubbers. The total capacity of retrofit scrubber installations is expected to be 30 GW for Phase I and 20 GW for Phase II. If we add to these figures the requirements for new power plants (from 30 to 100 GW) the market is going to be very large indeed.

4. The Supplier Industry

Inevitably markets as large as this create a large and powerful supplier industry. We will summarize the situation of the SO_2 and NO_x abatement industry in Japan and in the United States. Because of the substantial difference in emission standards, commercial applications of NO_x abatement technologies are widespread in Japan and much less so (with the exception of California) in the United States. About 100 plants in Japan are equipped with Selective Catalytic Reduction (SCR) NO_x control installations to achieve the required final concentration in the flue gas. SCR is generally used as a final step to remove the last 100 ppm of nitrogen oxides, after combustion control and use of low nitrogen fuels have provided the bulk of the action for low NO_x emission. The major suppliers of SCR systems are Babcock-Hitachi, Sumitomo Chemical, Hitachi Zosen, JGC Corporation, Japan Shell, Mitsubishi Heavy Industries (MHI), Ishikawajima-Harima Heavy Industries (IHI), and Mitsui Engineering and Shipbuilding.

Babcock-Hitachi has developed both pellet and plate type catalysts and installed about 30 SCR plants on various flue gases. Sumitomo Chemical uses granular catalysts in fixed and moving beds and its installations have been primarily on industrial boilers and heating furnaces at ammonia and methanol plants of the Sumitomo group. Hitachi Zosen uses honeycomb type catalysts called Noxnor. When the metallic honeycomb loses its reactibility because of contamination, it can be reactivated by washing. Hitachi has also developed an ammonia decomposition catalyst called Ammonon. The process marketed by JGC and Japan Shell is called Paranox, is based on a geometry of parallel plates and has found application in conditions of high ash content. MHI and IHI have shared the large utility market with catalysts acquired from producers under strict specifications and have to be regarded mainly as large systems suppliers.

In the United States, the most stringent NO_x regulations have been in California for refineries and petrochemical plants. The predominant system, with more than 25 applications, has been the Exxon thermal de-NO_x process. This selective non-catalytic reduction of NO_x with ammonia is based on the same principles and chemical reactions of the SCR processes, but without catalyst. The temperature is necessarily much higher (900 - 1000°C) and NH_3 is injected directly into the boiler. The process can achieve 50-60% NO_x reduction and is less expensive than SCR installations, but is also less efficient, harder to control and subject to corrosion and ammonium sulfates plugging.

The impending legislation is aimed at reducing the emissions by 2.5 million tons by the year 2001, which may be changed to 4.0 million tons by the Environmental Protection Agency (EPA) if deemed necessary for acid rain control and if proven cost effective. The emission rate limitations will be 0.45 pounds per million BTU for tangentially fired boilers and 0.5 pounds per million BTU for wall-fired boilers, unless those rates can not be achieved by low-NO_x burner technology. (Corresponding to 600-700 mg/m^3). Because of these relatively loose groundrules, one should not expect any substantial change to the NO_x abatement supplier industry in the United States.

The exception to this rule is represented by the very recent announcement that Southern California Edison will be installing an SCR system at a 480 MW unit in Redondo Beach. The suppliers are Noell and KRC (Knauf Research Cottrell).

Quite different is the forecast for Flue Gas Desulfurization. In preparation for the boom of retrofits and new plants expected after the passage of the Clean Air Act, the industry is undergoing some changes and, naturally, a large expansion. Babcock and Wilcox (B and W) and General Electric (General Electric Environmental Services or GEES) are probably the only major FGD companies which have not changed organization or ownership. All the other suppliers have undergone major changes. Research-Cottrell is now a division of Air and Water Technologies. Combustion Engineering and Peabody have now the same owner, Flakt, which is in turn is owned by Asea Brown Bovery. Wheelabrator is now offering the technology of Universal Oil Products (UOP) and has changed owner.

Newcomers in the field are Noxso, Union Carbide and General Atomics, all three offering advanced regenerative technologies: Noxso the adsorption process of the same name, Union Carbide the Cansolv liquid extraction process and finally General Atomics, through its joint venture with Ferlini Technology, the Ispra Mark 13A process.

Architect-engineering firms have also taken a direct position as suppliers of FGD technologies; most notable Bechtel that is now offering the Chiyoda technology. Other architect-engineers are taking a more direct approach in the business, including participation in the ownership of the plant. Some utilities are entering the supplier scene by acquiring air pollution control equipment firms (as in the case of San Diego Gas & Electric acquisition of

Wahlco and Bachmann) or offering engineering services (as in the case of Duke Power and its association with Fluor).

Also changing is the way in which the business is conducted. The old way of a utility acquiring a system from a technology supplier, hiring an architect-engineer to build it and then operating it will still be the predominant way, but new ideas are being circulated. One of these is offering the utilities a Flue Gas Cleaning service, in which ownership and operation of the scrubber remain with the supplier. Ferlini/General Atomics is offering the possibility of handling the marketing of the by-products sulfuric acid and hydrogen so that the utility will have no responsibility for it. A forecast on the technologies that will be preferentially chosen is very difficult. The rules of the game are still somewhat uncertain, but because of the emission caps and credits, there may be a substantial advantage for those suppliers offering technologies capable of removing the highest amount of SO_2 possible. And other questions are: for how long is the disposal of sludge going to be an economically viable alternative, and is gypsum going to be a tradeable commodity given the high impact of its transportation costs? Naturally, cost comparison will be the first factor determining the success or failure of a technology. The utility industry has provided comparative economic evaluations of different technologies through the Electric Power Research Institute (EPRI), the association of the majority of investor-owned utilities.

For several years, since its publication in 1985, the EPRI cost evaluation report [5], has represented a very useful guideline for the comparison of different systems. Early this year, the results of a new cost analysis for nine wet and six dry systems have been presented. The results of the analysis of new, regenerative systems are expected later this year. The "winner" in the wet category by a very small margin, in terms of levelized cost, has been the Pureair System, based on co-current limestone spraying with forced oxidation. Total levelized costs for this system range between 350 and 525 $/ton of SO_2 removed.

It should be noted that the "worst" wet system shows a range of costs between 400 and 600 $/ton SO_2 removed, with all the others in between, which indicates the high competitiveness of the industry. In the dry injection field the best calculated numbers belong to the Lurgi CFB System with 330 to 510 $/ton SO_2 removed. In this case, the spread is higher with the worst case being 520 to 770 $/ton removed. The new EPRI estimates are based on a 300 MW plant burning medium-sulfur (2.6%) coal. Plant size and coal characteristics are in line with the majority of the coal-fired boilers requiring scrubbers as a result of Clean Air Act.

In Japan, Flue Gas Desulfurization Plants accounts for 18 GW of power capacity (as of March 1989), equally divided between coal and oil. The summary of the Japanese installations is given in Table 2.

TABLE 2
FGD INSTALLATION IN JAPAN FOR POWER UTILITY COMPANIES

I. Capacity	
Total	18,023.57 MW (100%)
Oil	8,863.75 MW (49%)
Coal	9,159.82 MW (51%)
II. Process	
Wet Lime/Limestone-Gypsum	15,997.57 MW (88.7%)
Oil	7,965.99 MW (44.2%)
Coal	8,031.58 MW (44.5%)
Dual Alkali	1,400 MW (7.7%)
Wellman Lord	200 MW (1.2%)
Mg-Gypsum	466 MW (2.2%)

Almost all the lime/limestone plants installed at coal burning power stations produce salable gypsum. The most common process was developed by Mitsubishi Heavy Industries (MHI), characterized by a water pre-scrubber and by pH adjustment with sulfuric acid. Other processes using a variety of additives to reduce scaling are: the Kawasaki Heavy Industries (KHI) magnesium-gypsum, the Dowa's aluminum sulfate-limestone, Kureha Chemical Industries' sodium acetate-lime-gypsum, and Kobe Steel's calcium chloride lime-gypsum.

A second generation process is the Chiyoda CT-121. The process is based on a unique absorber, known as the jet bubbling reactor (JBR) in which SO_2 removal and oxidation to sulfate take place simultaneously. In the reactor, flue gas is first water prescrubbed and then is injected into a limestone-gypsum slurry from submerged pipes, and air is blown directly into the slurry for oxidation. The schematic of the process, in the American Bechtel version, is shown in Figure 9.

Construction of the first CT-121 unit was completed in 1984 at the Toyama Joint Power's 400 MW power plant. Since then, extensive tests have been carried out at the Takehara plant and a new design, without prescrubber has been evaluated.

Babcock-Hitachi has developed a new FGD system in which prescrubbing, SO_2 absorption and calcium sulfite oxidation are combined in a single tower. The company is calling this FGD system the "Intelligent Type". A schematic comparison with a conventional FGD is shown in Figure 10.

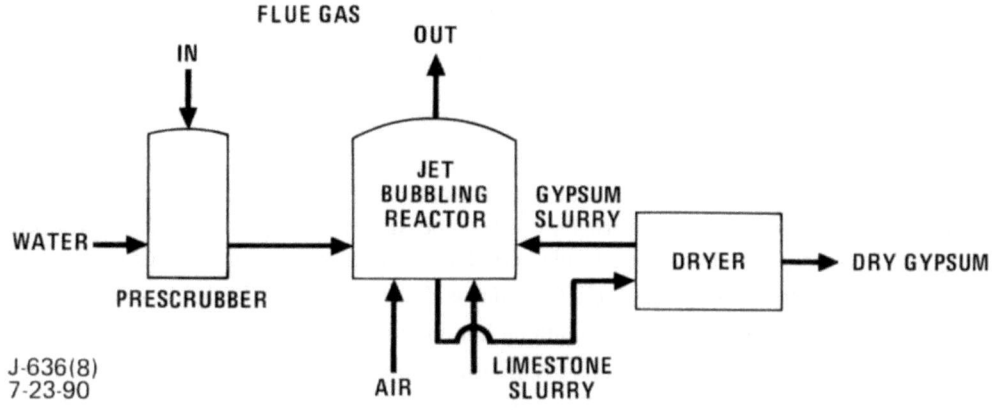

Figure 9. Bechtel CT-121 simplified process flow schematic

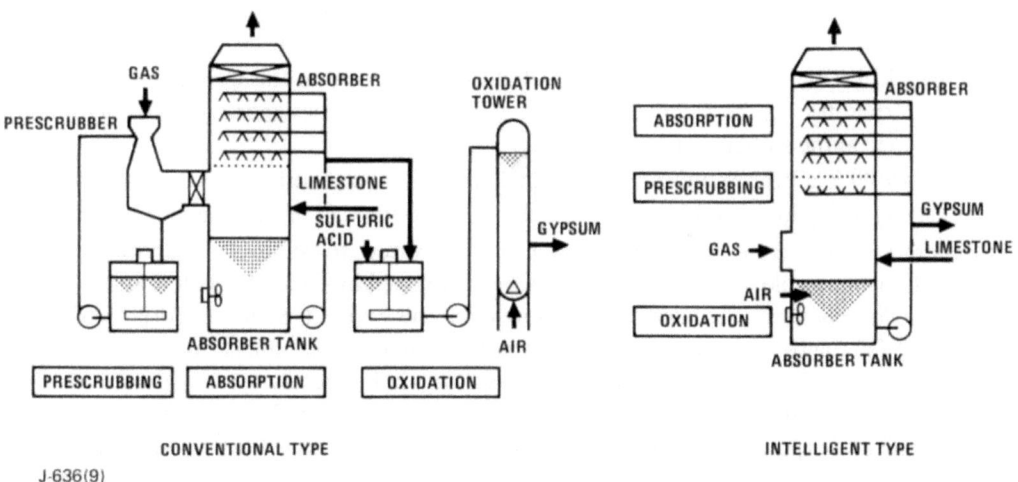

Figure 10. Comparison of systems

5. Emerging Technologies in the United States

Several new processes for SO_2 and NO_2 abatement have been announced in the media by various industrial suppliers. Neglecting the chemical and engineering modifications of the lime/limestone process, we selected three technologies that seem to have attracted the most attention in the specialized literature: Noxso, Cansolv and Ispra.

Noxso [8] is a dry absorption process in which the flue gas is cleaned in a fluidized bed of a solid sorbent (sodium carbonate coated aluminum oxide) at 120°C. Spent sorbent is regenerated by treatment with hydrogen or methane at 600°C. The regeneration produces SO_2 and H_2S which are converted to elemental sulfur in a conventional Claus plant. The fluidized bed reactor also absorbs the NO_x which is then desorbed with hot air at 600°C prior to sorbent regeneration. The NO_x containing hot air is mixed with the combustion air fed to the burners. The recycled NO_x suppresses the formation of new NO_x during combustion and eventually a steady state is achieved with relatively low NO_x emission.

The schematic of the process is shown in Figure 11. The process has been tested at the laboratory size level (fluidized bed reactor of 1 meter in diameter and 1 meter in height and sorbent regeneration reactor of 0.6 meters in diameter). Scale-up tests are planned at an Ohio Edison power plant (Toronto Station) for an equivalent 5 MW unit. A full-scale, two-year demonstration of the technology is also scheduled at another Ohio Edison plant (Niles Station) with construction scheduled to begin in late 1991, as part of the Department of Energy Clean Coal Technology Program. Industries involved in the development are the Noxso Corporation as the technology owner, MK-Ferguson as designer and constructor and W. R. Grace as the sorbent supplier.

Cansolv [9] is a process utilizing an organic amine as the absorbent in a countercurrent multi-stage Waterloo Scrubber. The SO_2 is recovered from the amine by thermal regeneration and then marketed as such or converted to sulfur or sulfuric acid by conventional technologies. The process is being developed by Union Carbide Canada and a 2 MW pilot plant is scheduled for start-up later this year at Suncor's Fort McMurray (Alberta) Oil Sands Plant, processing a flue gas slipstream from boilers burning 7% sulfur petroleum coke. The schematic of the process is shown in Figure 12.

Through the formation of the joint venture company Ferlini/General Atomic Development Corporation (FGD), the Mark 13A process developed by the Joint Research Centre Ispra of the European Community is being marketed in the United States and in Japan under the name ISPRA process.

The process remains unique amongst regenerative process for its single stage absorption-oxidation of SO_2 to sulfuric acid. The American development efforts have been primarily directed at engineering flow sheet optimization and at cost optimization. The process has been included in the comparative cost analysis that EPRI is conducting on conventional and new technologies. While more advanced in development than other new technologies also the ISPRA process will probably not be commercially available for Phase 1 of the Clean Air

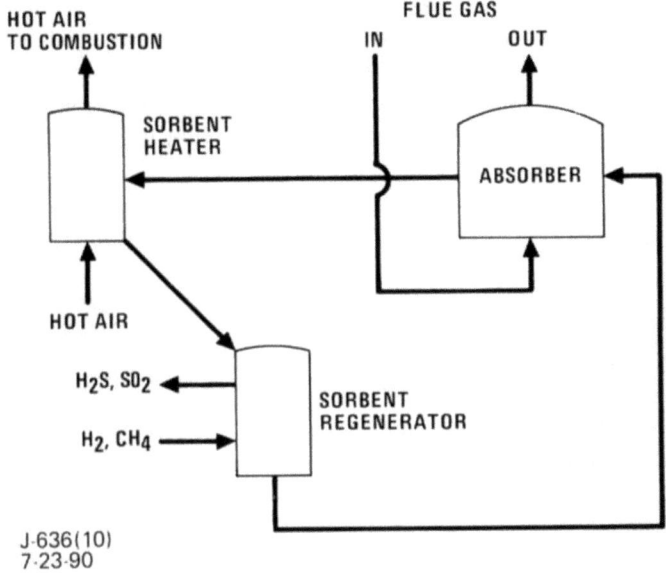

Figure 11. Noxso process schematic

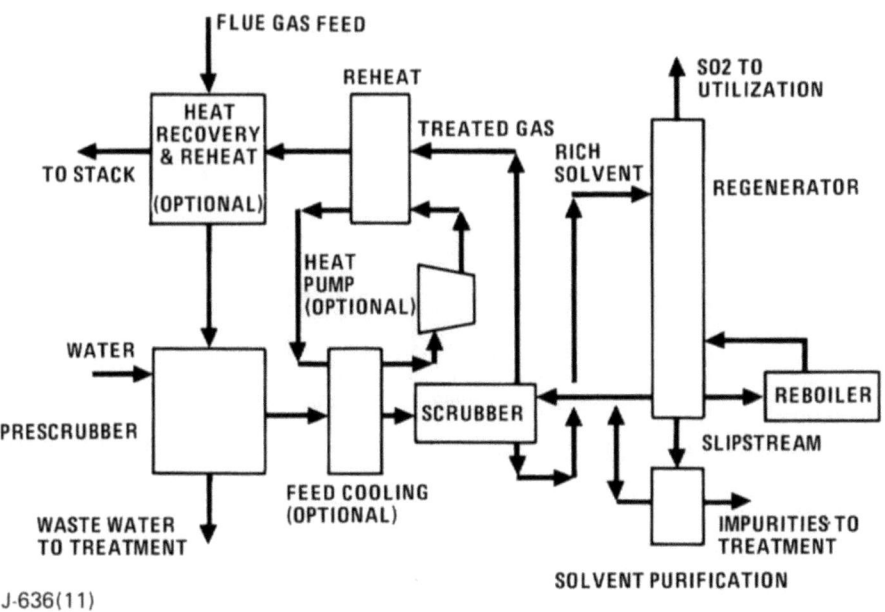

Figure 12. Cansolv process schemmatic

Act, but it can represent a viable alternative to utilities for compliance for Phase 2, in which the systems will have to be ordered by 1996-1997.

At this point in time, given the uncertainties in the regulatory and political situation, a prediction on the success or failure of the conventional and new technologies is not possible. What is certain is that the American market for FGD system is going to offer great opportunities to all the participants as it is shown, as a conclusion, in Figure 13.

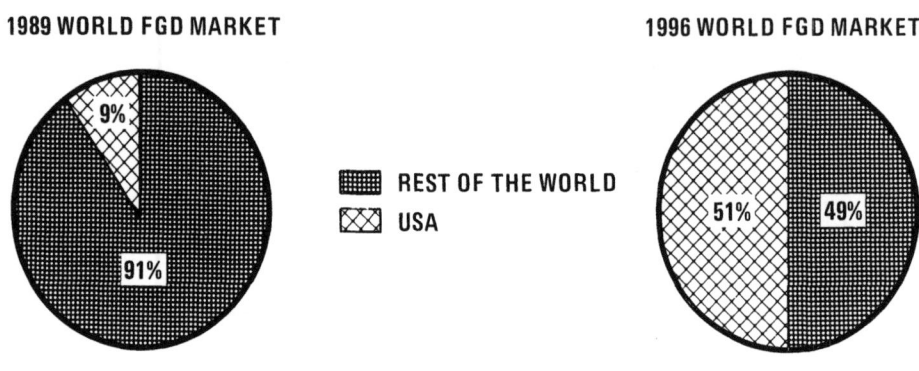

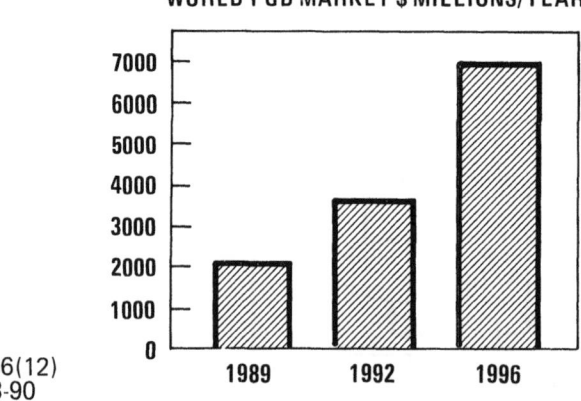

Figure 13. World FGD market

References

1) Miyahara, S. (1990), Gen-Upgrade 90, Washington, D.C.

2) Inventory of Power Plants in USA (1988) Energy Information Administration, Washington, D.C.

3) Noyes, R. (ed.), (1978) Coal Resourses in USA, Noyes Data Corp., Park Ridge, NJ

4) FGD Newsletter (1990), McIlvaine Co., Northbrook, IL

5) Keeth, R.J. and Ireland, P.A. (1985) Economic Evaluation of FGD Systems, EPRI Report CS-3342, Palo Alto, CA

6) Keeth, R.J., Ireland, P.A. and Radcliffe, P.T., (1990) Upgrade of FGD Economic Evaluations, SO_2 Control Symposium, New Orleans, Louisiana

7) Goto, O. and Ikeda, S. (1990) The Status and Prospect of FGD on Thermal Power Plant in Japan, SO_2 Control Symposium, New Orleans, Louisiana.

8) Haslbeck, J.L., Neal, L.G. and Ma, W.T. (1988) Modern Power Systems, July 1988

9) Hakka, L.E. and Burgess, J.S. (1990) The Cansolv FGD Process, SO_2 Control Symposium, New Orleans, Louisiana

If you have any concerns about our products,
you can contact us on
ProductSafety@springernature.com

In case Publisher is established outside the EU,
the EU authorized representative is:
**Springer Nature Customer Service Center GmbH
Europaplatz 3, 69115 Heidelberg, Germany**

Printed by Libri Plureos GmbH
in Hamburg, Germany